# A to Z
# Math Games

by Karen M. Breitbart
illustrated by Marilynn G. Barr

A special thanks to . . .

❤ The Citibank Success Fund for providing the grant that helped to purchase the items necessary for *A to Z Math Games*.
❤ Mr. Robert F. Morgan, my principal, for allowing me the freedom to try new things in my classroom.
❤ Most importantly, my loving husband, Gregg, for always encouraging me to succeed.

## About the Author:

Karen Breitbart received a Bachelor of Science degree in Education at Louisiana State University and a Master's degree in Early Childhood Education at Florida International University. She has been teaching for ten years.

## Other Books of Interest:

*A to Z Language Games*, which incorporates games and activities to strengthen language skills and concepts.

**Publisher:** Roberta Suid
**Educational Consultant:** Lillian Lieberman
**Editor:** Hawkeye McMorrow
**Design & Production:** Scott McMorrow
**Cover Art:** Marilynn G. Barr

For a complete catalog, please write to the address below:
P.O. Box 1680, Palo Alto, CA 94302

Please visit our web site: http//www.mondaymorningbooks.com

or e-mail us at MMBooks@aol.com

Monday Morning is a registered trademark of
Monday Morning Books, Inc.

ISBN 1-57612-016-3
Printed in the United States of America
98765432

# Contents

# Introduction

*A to Z Math Games* reinforces basic math skills and concepts. The games in this resource will help children build a strong math foundation on which to base future learning experiences.

The *A to Z Math Games* program assists children in achieving success by providing them with tools for hands-on exploration and manipulation. These high-interest, self-motivating games are intended for individual use.

All games in this resource have been "kid-tested" and approved! The children loved to learn this way and looked forward to learning math skills. Each child was able to work at his or her own pace, and they progressed, regardless of their varying abilities.

After you have made the games, you will have them for years. Since only one child plays with each game at a time, you avoid the problem of losing parts.

I hope that you and your students enjoy these games as much as my students and I enjoyed creating them!

## IN THE CLASSROOM

A good technique for incorporating *A to Z Math Games* in the classroom is to set up a Math Center containing three to five of the games per week. There are many ways to organize the use of these games. Here is one possibility for using five games per week:

## Making the Games:

- Each game contains directions for making and using motivating hands-on games in a classroom setting.
- For each game, all patterns are included. Some games may require additional readily available materials.
- Color the patterns as desired.
- To keep your games in the best possible shape, laminate the pieces. If you do not have access to a laminating machine, you can also use clear contact paper to cover the pieces.
- Consider labeling patterns on the back for children to self-check their work.

## Setting up a Math Center Area:

- Find five different colored plastic baskets (or use five of the same colored baskets, labeled with different colored construction paper).
- Label the table you've chosen for your center area with five construction paper circles. These circles should be the same color as the baskets or the basket labels.
- Permanently attach these circles to the table using tape and then cover them with clear Contac paper. The baskets will always be stored on top of the corresponding colored circle.
- Select five games that you would like your students to work with. Put one game in each basket.

## Preparing Math Center Groups:

- Divide your children into groups of five.
- Assign each child a color, so that there is one "red," "yellow," "blue," "orange," and "green" child in each group. (Your colors may vary, depending on what color baskets you use.)
- Set up your groups before the children come in to school and write the children's names on the colored circles that are permanently attached to the Math Center table. (Write the names with a permanent marker.)
- Allow the children to sit with their groups and decide on a group name. These children will stay together all year.
- Tell the children what their colors are.

 A to Z Math Games © 1997 Monday Morning Books, Inc

## How to Use the Games:

• On Monday, introduce these games to the children. Directions are included for you to read to the children.

• First, model each game step-by-step, from setting up to playing to cleaning up.

• After you have modeled each game, choose one or two children to demonstrate how to set up, play with, and clean up the game.

• At Center Time, have the groups rotate through all of the centers in your room, including the Math Center. When a group gets to the Math Center, each child will take and use the game that is in his or her color basket.

• At Clean Up Time, children will replace their game in the basket and place it on the corresponding color, making sure the games are ready for the next group.

• At the end of the day, rotate the games from left to right. The game in the red basket will move to the blue basket, the blue to the orange, and so on, with the last game in the row becoming the first. By rotating the games, the children will find a different game in their basket each day and you will be able to keep track of who is playing which game. By the end of the week, all of your children will have had time to play with each game.

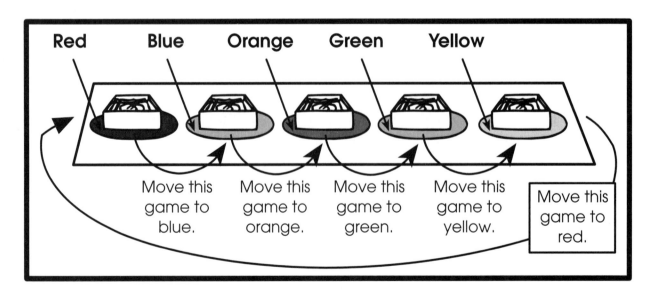

# HUNGRY ALLIGATORS

## Objective:
• Children will make number sets from 1 to 5.

## Materials:
Alligators (p. 8), Fish (p. 9), scissors, five brads, hole punch, crayons or markers, resealable bag

## How to Make the Game:
1. Duplicate the Alligators and Fish, color, laminate, and cut out.
2. Punch a hole in each of the Alligators and in the Alligators' mouths (as indicated).
3. Attach the mouths to the Alligators using brads.
4. Store all game pieces in the resealable bag.

## How to Play the Game:
1. Take the Alligators and Fish out of the bag.
2. Each Alligator has a number on its hat. Place the Alligators in a line from 1 to 5.
3. Match sets of the Fish to the number on each Alligator. For example, for the first Alligator, you will have a set of one Fish.
4. "Feed" the Fish to the Alligators.
5. When you are finished, have your teacher check your work. Then place all game pieces back in the bag.

## Book Link:
• *Alligators All Around* by Maurice Sendak (Harper, 1962). Alligators perform activities for each letter of the alphabet.

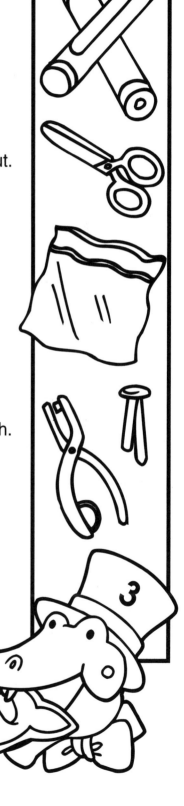

A to Z Math Games © 1997 Monday Morning Books, Inc.

# Alligators

# Fish

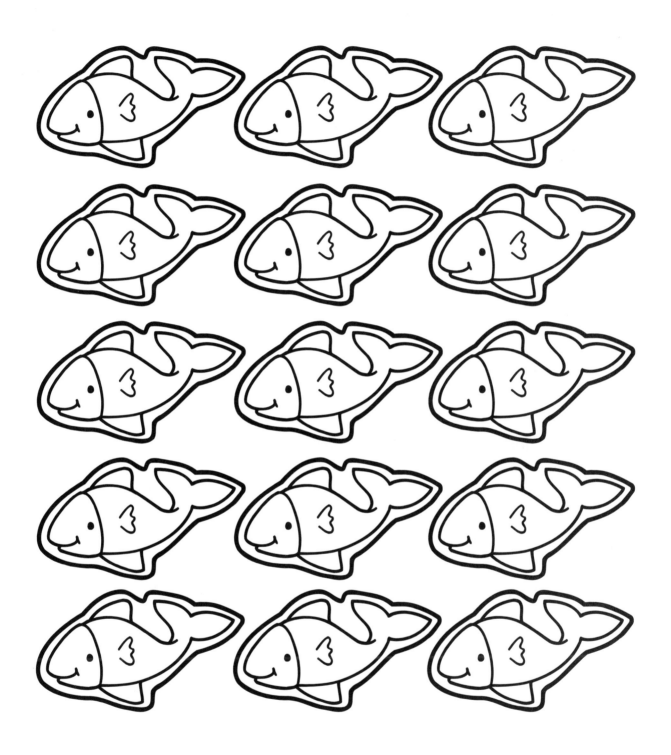

 A to Z Math Games © 1997 Monday Morning Books, Inc.

# AIRPLANE MATH

## Objective:
- Children will match dots to the correct numerals.

## Materials:
Airplane Bodies (p. 11), Airplane Tails (p. 12), crayons or markers, scissors, resealable bag

## How to Make the Game:
1. Duplicate the Airplane Bodies and Tails, color, laminate, and cut out.
2. Store all game pieces in the resealable bag.

## How to Play the Game:
1. Take the game pieces out of the bag. Line the Airplane Bodies in one row and the Airplane Tails in another.
2. The Airplane Bodies have dots, and the Tails have numbers.
3. Place each Airplane Tail on the correct Airplane Body.
4. When you are finished, have your teacher check your work. Then place all game pieces back in the bag.

## Book Link:
- *Going on an Airplane* by Fred Rogers, photographs by Jim Judkis (Putnam, 1989).

This book describes a plane trip, from packing to arriving.

# Airplane Bodies

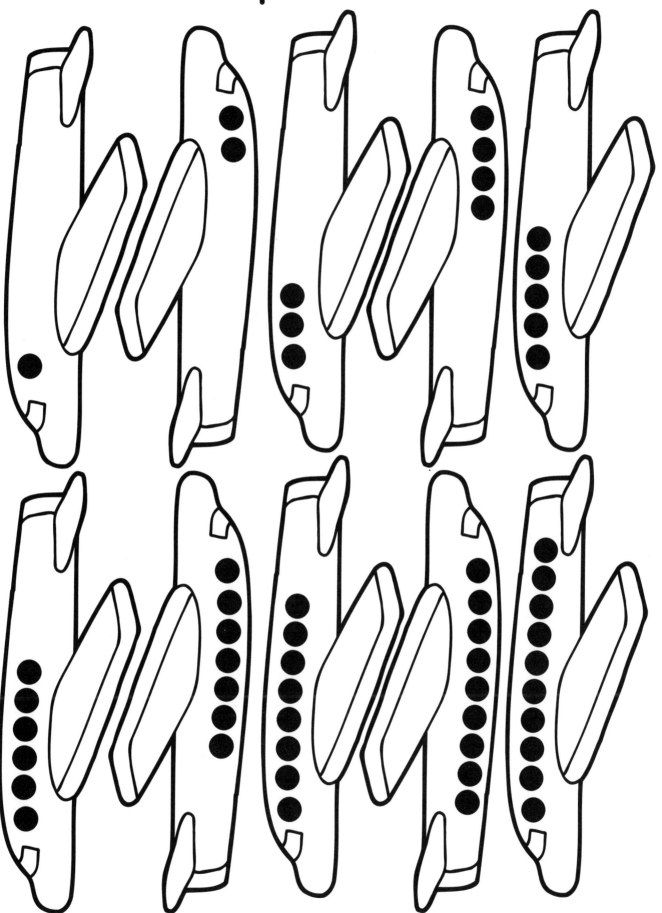

A to Z Math Games © 1997 Monday Morning Books, Inc.

# Airplane Tails

# BEAN TOSS

## Objective:
- Children will count and record numbers on a graph.

## Materials:
Happy Bean Graph (p. 14), crayon, permanent marker, five large lima beans, large envelope, resealable bag

## How to Make the Game:
1. Duplicate the Happy Bean Graph. Make one copy per child.
2. Use the permanent marker to draw a happy face on one side of each lima bean.
3. Store the beans and a crayon in the resealable bag, and store the Happy Bean Graphs in a large envelope.

## How to Play the Game:
1. Take one Happy Bean Graph out of the envelope.
2. Take the beans out of the bag and toss them onto the table.
3. Sort the beans into two piles: one with the happy faces up and one with the blank sides up.
4. Count the number of happy faces and record this on the graph by writing the numeral in the correct column. Toss the beans four times.
5. When you are finished, give your graph to your teacher. Then place the beans and the crayon in the resealable bag.

## Book Link:
- *Jack and the Beanstalk* by Steven Kellogg (Scholastic, 1965). Kellogg retells the story in black and white with green additions.

13    A to Z Math Games © 1997 Monday Morning Books, Inc.

# Happy Bean Graph

## How many?

### HAPPY BEAN TOSS

NAME _____

| 0 | 1 | 2 | 3 | 4 | 5 |
|---|---|---|---|---|---|
|   |   |   |   |   |   |
|   |   |   |   |   |   |
|   |   |   |   |   |   |
|   |   |   |   |   |   |
|   |   |   |   |   |   |

# BLOCK NUMBERS

## Objective:

- Children will use blocks to make number sets from 1-10.

## Materials:

Number Cards (p. 16), 55 small wooden blocks or beads, scissors, two large resealable bags

## How to Make the Game:

1. Duplicate the Number Cards, laminate, and cut apart.
2. Store the Number Cards in one bag and the blocks in another.

## How to Play the Game:

1. Line the Number Cards in order from one to ten.
2. Place the correct number of blocks in front of each card. For example, place one block by the card that has the number 1.
3. When you are finished, have your teacher check your work. Then place the blocks back in one bag and the Number Cards in the other.

## Option:

Children can use blocks to outline the numbers on enlarged Number Cards.

## Book Link:

- *Hippos Go Berserk! (a silly sort of counting book)* by Sandra Boynton (Recycled Paper Press, 1977).

This counting book will definitely make children smile.

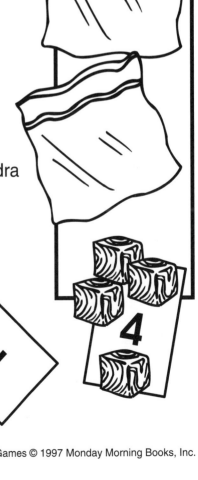

# Number Cards

| | |
|:---:|:---:|
| 5 | 10 |
| 4 | 9 |
| 3 | 8 |
| 2 | 7 |
| 1 | 6 |

# CALICO CATS

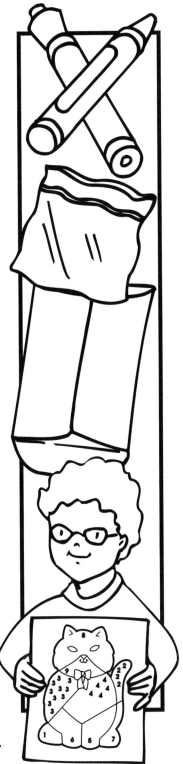

## Objective:
• Children will practice writing the numbers 1-9.

## Materials:
Cat (p. 18), nine different colors of crayons or markers, resealable bag, large envelope

## How to Make the Game:
1. Duplicate one copy of the Cat per child.
2. Store the Cats in a large envelope and the crayons or markers in the resealable bag.

## How to Play the Game:
1. Take one copy of the Cat out of the envelope.
2. Find the small numbers in each section of the Cat.
3. Use a different color crayon or marker to write each number over and over in each section.
4. When you are finished, place the crayons or markers back in the bag, and give your picture to the teacher.

## Note:
Post the finished Cat pictures on a "Calico Cat" bulletin board.

## Variation:
Use masking tape labels to number the crayons from one to nine. Children can color their Cats using the numbered crayons.

## Book Link:
• *The Antique Store Cat* by Leslie Baker (Little, Brown, 1992). A calico cat makes an "unauthorized" visit to an antique store.

 A to Z Math Games © 1997 Monday Morning Books, Inc.

# Cat

# CLOCKS

## Objective:
- Children will practice telling time.

## Materials:
Clock and Clock Hands (p. 20), Clock Cards (p. 21), heavy paper, brad, crayons or markers, hole punch, scissors, large envelope, resealable bag

## How to Make the Game:
1. Duplicate the Clock and Clock Hands onto heavy paper, color, laminate, and cut out.
2. Punch a hole in the center of the Clock, and attach the Clock Hands using the brad.
3. Duplicate the Clock Cards, color, laminate, and cut out.
4. Store the Clock in a large envelope and the Clock Cards in the resealable bag.

## How to Play the Game:
1. Take the Clock out of the envelope and point the hands to 12.
2. Take the Clock Cards out of the bag and place them in a row.
3. One by one, look at the clock on each Clock Card. Then move the hands on the Clock so that they are the same as the Clock on each card. Read the times aloud.
4. When you are finished, place the Clock back in the envelope and place the Clock Cards in the bag.

## Book Link:
- *The Grouchy Ladybug* by Eric Carle (Crowell, 1977). A grouchy ladybug goes through a whole day looking for a fight.

 A to Z Math Games © 1997 Monday Morning Books, Inc.

# Clock and Clock Hands

# Clock Cards

A to Z Math Games © 1997 Monday Morning Books, Inc.

# DOMINO NUMBERS

## Objective:
- Children will use the dominoes to form the numbers 1 to 10.

## Materials:
10 pieces of tag board, crayons or markers, dominoes, large envelope, resealable bag

## How to Make the Game:
1. Write one number (from 1 to 10) on each piece of tag board.
2. Store the number cards in the large envelope and the dominoes in the resealable bag.

## How to Play the Game:
1. Spread the number cards face up on the floor or on a table.
2. Use the dominoes to trace the numbers. Match the dots on the first domino with another domino piece that has the same number of dots. Continue matching as you trace the number.
3. When you have traced the numbers on all of the cards, have your teacher check your work. Then place the cards in the envelope and the dominoes in the bag.

# IN THE DOGHOUSE

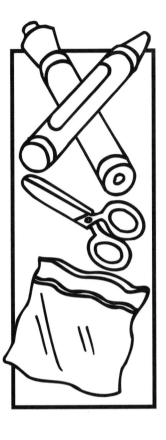

## Objective:
• Children will match dots to numerals.

## Materials:
Dalmatians (p. 24), Doghouses (p. 25), crayons or markers, scissors, resealable bag

## How to Make the Game:
1. Duplicate the Dalmatians, laminate, and cut out.
2. Duplicate the Doghouses, color, laminate, and cut out.
3. Store all game pieces in the resealable bag.

## How to Play the Game:
1. Take the Doghouses and Dalmatians out of the bag.
2. Each Doghouse has a number on it. Line the Doghouses in a row from 1 to 10.
3. Count the spots on each Dalmatian.
4. Match each Dalmatian to the correctly numbered Doghouse.
5. When you are finished, have your teacher check your work. Then place all game pieces back in the bag.

## Book Link:
• *The Dog Who Thought He Was a Boy* by Cora Annett (Houghton Mifflin, 1965).
The Poppersons try to make Ralph understand that he's a dog.

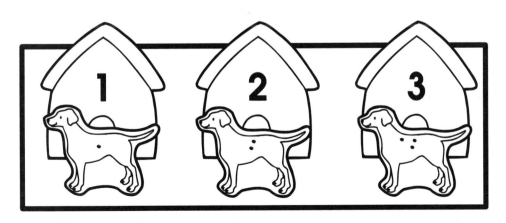

# Dalmatians

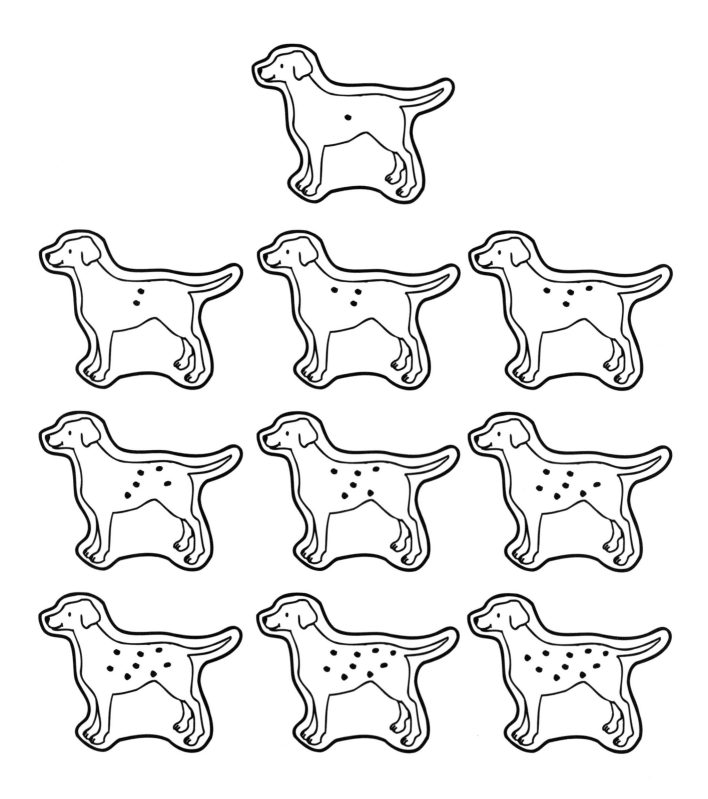

# Doghouses

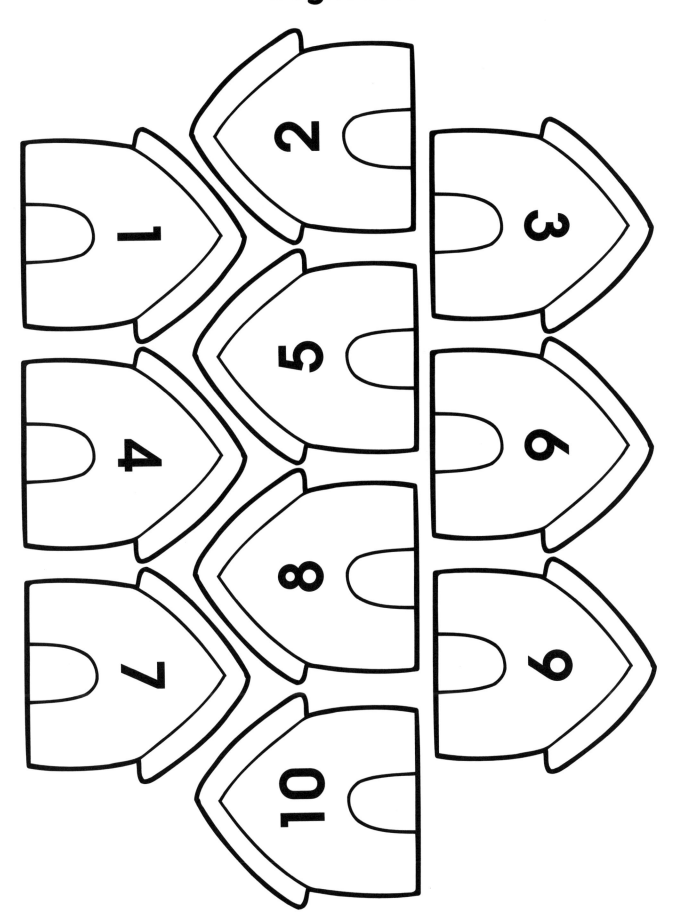

 A to Z Math Games © 1997 Monday Morning Books, Inc.

# EGGS ARE HATCHING

## Objective:

• Children will use counters to solve addition problems.

## Materials:

Eggs (p. 27), Chicks (p. 28), crayons or markers, scissors, hole punch, 6 brads, counters, resealable bag

## How to Make the Game:

1. Duplicate the Eggs, laminate, and cut out.
2. Duplicate the Chicks, color, laminate, and cut out.
3. Punch a hole in each Egg and each Chick (as indicated).
4. Use a brad to attach each Chick to the back of the correct Egg.
5. Store all game pieces and counters in a resealable bag.

## How to Play the Game:

1. Spread the Eggs in a row in front of you.
2. Use the counters to solve the problem on each Egg.
3. Check your work by sliding the Chick up from behind each Egg. The number on the Chick should match your answer.
4. When you are finished, place all game pieces in the bag.

## Option:

Don't attach the Chicks to the Eggs. Have children match the correct Chick to each Egg.

## Book Link:

• *The Amazing Egg Book* by Margaret Griffin and Deborah Seed, illustrated by Linda Hedry (Addison-Wesley, 1989).
This book discusses where eggs come from and much more!

# Eggs

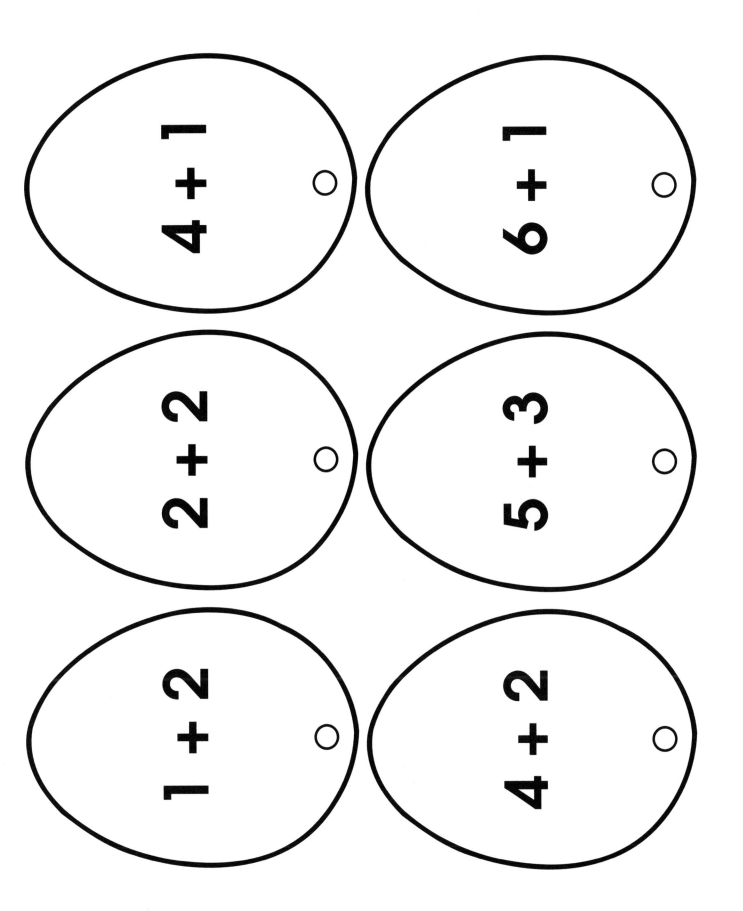

 A to Z Math Games © 1997 Monday Morning Books, Inc.

# Chicks

# EGG CARTON GAME

## Objective:
• Children will practice writing the numbers 1-6.

## Materials:
Record Sheet (p. 30), six-sectioned egg carton, permanent marker, crayon, marble, large envelope, resealable bag

## How to Make the Game:
1. Duplicate one copy of the Record Sheet per child.
2. Write a number from 1 to 6 in each section of the egg carton.
3. Store the crayon and marble in the bag, inside the carton.
4. Store the Record Sheets in the large envelope.

## How to Play the Game:
1. Take one Record Sheet out of the envelope, and take the marble and crayon out of the bag.
2. Place the marble in the egg carton and close the top.
3. Shake the egg carton with the marble inside.
4. Open the egg carton to see where the marble landed.
5. Record the number of the egg carton section in the correct row by writing the number on one of the egg patterns.
6. Continue shaking the marble and marking the number on the Record Sheet until you fill up one row.
7. When you are finished, give your Record Sheet to your teacher. Then place the marble and the crayon in the bag and the bag inside the egg carton.

## Option:
Have children compare their finished sheets. Make a class bar graph showing which number "won."

## Book Link:
• *A Caribbean Counting Book* (Houghton Mifflin, 1996). This is a delightful collection of rhymes to be chanted.

 A to Z Math Games © 1997 Monday Morning Books, Inc.

# Record Sheet

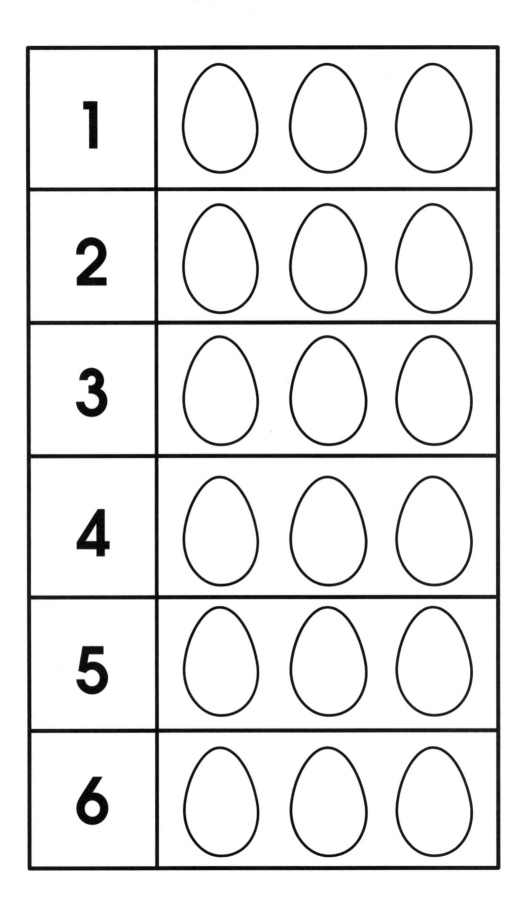

# FISH GAME

## Objective:
- Children will make sets for the numbers 1 to 10.

## Materials:
Number Fish (p. 32), crayons or markers, small magnet with a hole in the center, ruler or stick, yarn, ten paper clips, scissors, counters, resealable bag

## How to Make the Game:
1. Duplicate the Number Fish, color, laminate, and cut out.
2. Attach a paper clip to each Number Fish.
3. Attach the length of yarn to the end of the ruler or stick.
4. Tie the magnet to the end of the piece of yarn.
5. Store the Number Fish and counters in the resealable bag.

## How to Play the Game:
1. Spread the Number Fish on the floor.
2. Go fishing with the fishing pole!
3. After catching a fish, look at the number on the front of it.
4. Count out the same amount of counters and place them next to the fish.
5. After catching all of the fish, place the Number Fish and the counters back in the bag.

## Book Link:
- *One Fish, Two Fish, Red Fish, Blue Fish* by Dr. Seuss (Random House, 1960).

This rhyming classic is perfectly suited for this counting game.

 A to Z Math Games © 1997 Monday Morning Books, Inc.

# Number Fish

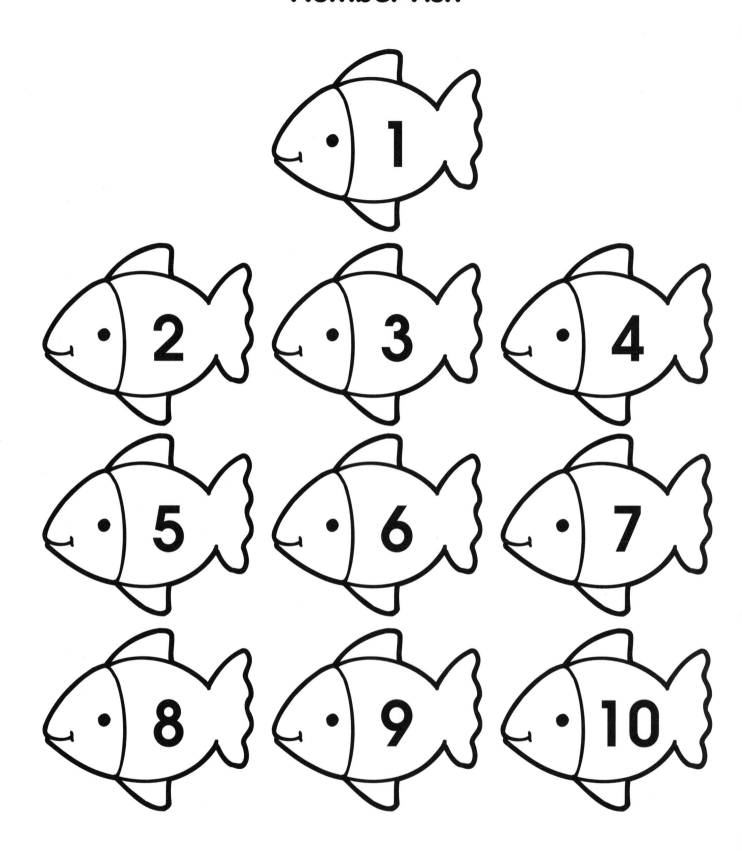

# FROG GRAPH

## Objective:
• Children will practice sorting, counting, and graphing.

## Materials:
Frogs (p. 34), Frog Graph (p. 35), crayons or markers, scissors, large envelope, resealable bag

## How to Make the Game:
1. Duplicate the Frogs, color, laminate, and cut out. Make as many copies of the frogs as you'd like, up to ten of each.
2. Duplicate one copy of the Frog Graph per child.
3. Store the Frog Graphs in the large envelope and the Frogs in the resealable bag along with one crayon or marker.

## How to Play the Game:
1. Take one copy of the Frog Graph out of the envelope.
2. Take the Frogs out of the bag and divide them by patterns: polka dots, stars, and stripes.
3. Use the crayon to mark the total number of each type of Frog on the Frog Graph.
4. When you are finished, place the crayon and the Frogs in the bag and give your graph to your teacher.

## Option:
Bind the Frog Graphs in a classroom Graph Book.

## Book Link:
• *Big Frog, Little Pond* by George Mendoza (McCall, 1971).
A big frog disturbs his neighbors with his loud croaking.

# Frogs

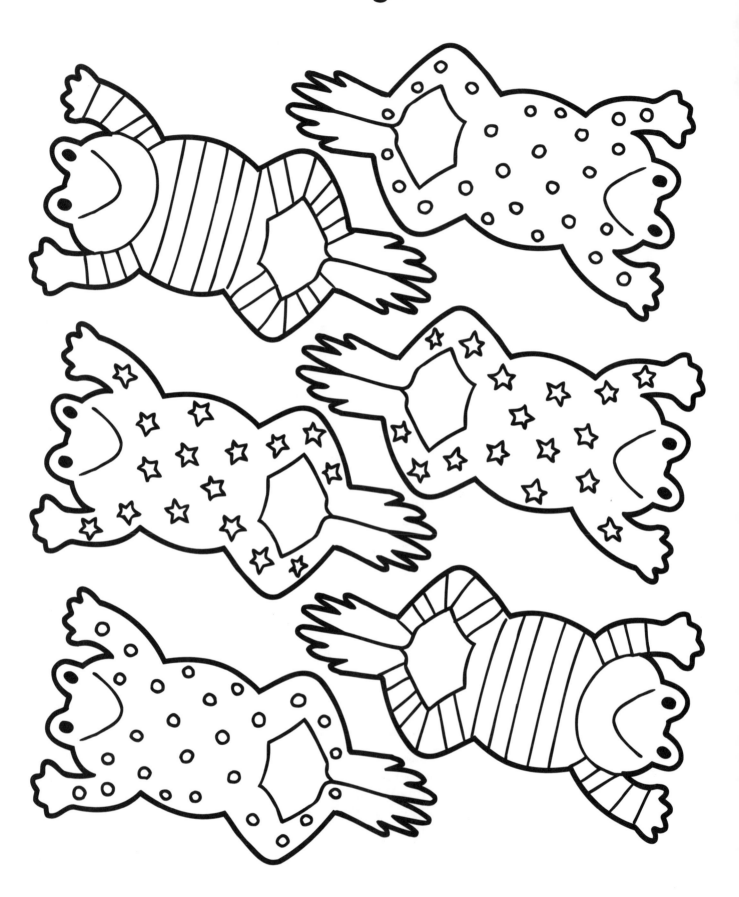

# Frog Graph

| | | | |
|---|---|---|---|
| 10 | | | |
| 9 | | | |
| 8 | | | |
| 7 | | | |
| 6 | | | |
| 5 | | | |
| 4 | | | |
| 3 | | | |
| 2 | | | |
| 1 | | | |
| | | | |

 A to Z Math Games © 1997 Monday Morning Books, Inc.

# GHOST PUZZLE

## Objective:
• Children will practice matching dots to numbers.

## Materials:
Ghost (p. 37), dice, scissors, resealable bag

## How to Make the Game:
1. Duplicate the Ghost, laminate it, cut it out, and cut it apart along the puzzle lines.
2. Store all game pieces and the dice in the resealable bag.

## How to Play the Game:
1. Take the pieces out of the bag and put the puzzle together.
2. Roll the dice and take away the piece of the puzzle that has the same number on it as the number of dots on the dice.
3. Continue to roll and take away pieces until all parts of the Ghost are gone. (If you roll a number you've already removed, roll again.)
4. When you are finished, place all game pieces back in the bag.

## Note:
This game can also be played with a spinner. Each section of the spinner should have a dot on it (from one to ten).

## Book Link:
• *Gus Was a Real Dumb Ghost* by Jane Thayer (Morrow, 1982).
A ghost decides to go to school and learn to spell when a publisher returns his autobiography.

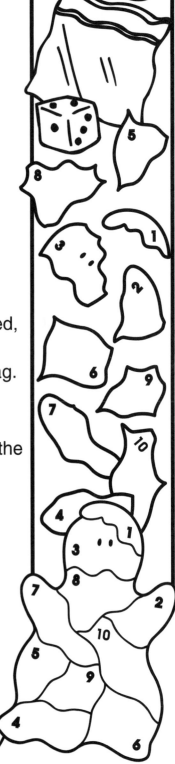

# Ghost

# GOING FOR GOLD

## Objective:
- Children will solve simple addition problems using counters.

## Materials:
Gold Coins (p. 39), Treasure Chests (p. 40), crayons or markers (including gold), scissors, counters, resealable bag

## How to Make the Game:
1. Duplicate the Gold Coins, color gold, laminate, and cut out.
2. Duplicate the Treasure Chests, color, laminate, and cut out.
3. Store all game pieces and counters in a resealable bag.

## How to Play the Game:
1. Spread the game pieces face up on a table. Line the Treasure Chests in a row and keep the Gold Coins.
2. Solve the addition problem on each Gold Coin. (Use counters if you need help.)
3. Match each Gold Coin with the Treasure Chest that has the correct answer.
4. When you are finished, have your teacher check your work. Then place all game pieces back in the bag.

## Book Link:
- *Clever Tom and the Leprechaun* retold and illustrated by Linda Shute (Scholastic, 1988).

All Tom has to do to get the leprechaun's gold is catch him. But this isn't as easy as it sounds.

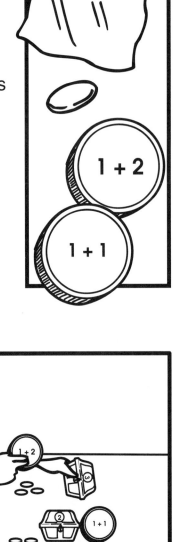

# Gold Coins

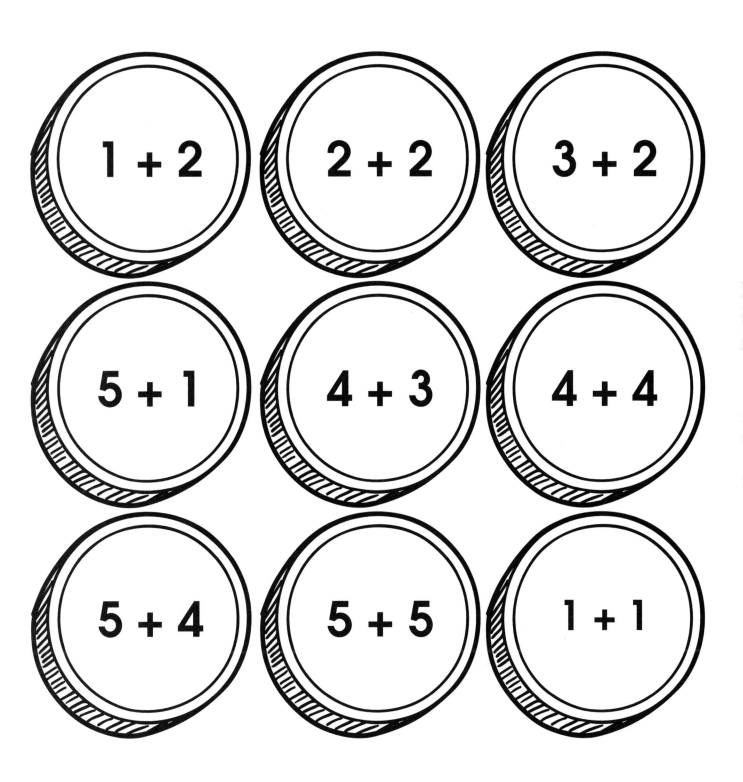

A to Z Math Games © 1997 Monday Morning Books, Inc.

# Treasure Chests

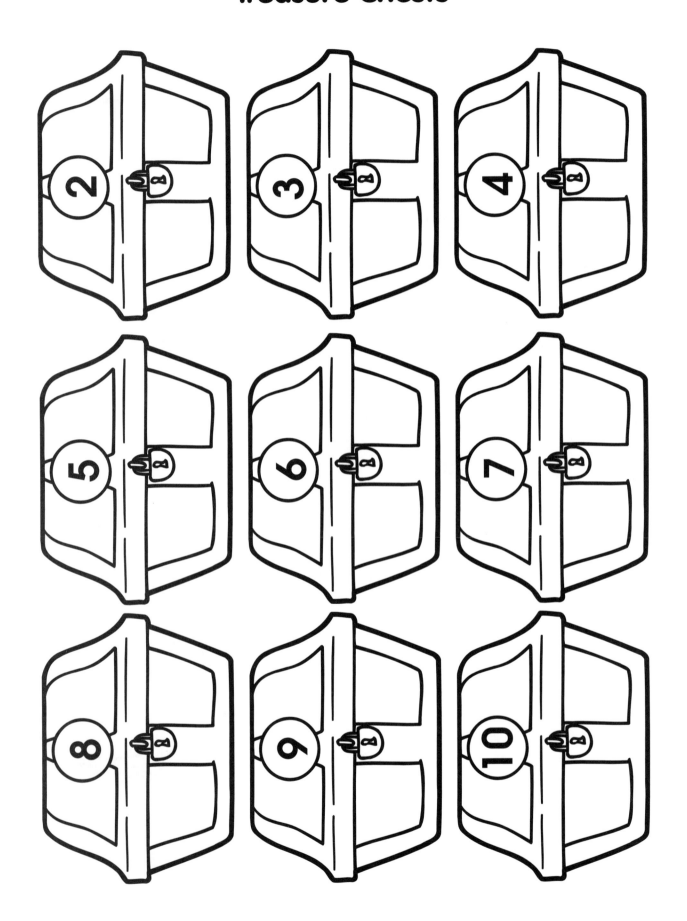

# HALLOWEEN MATH

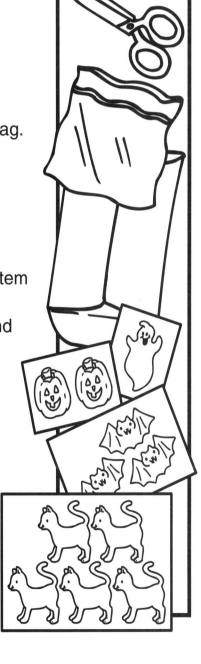

## Objective:
• Children will practice counting and graphing.

## Materials:
Halloween Counters (p. 42), Halloween Graph (p. 43), crayons or markers, scissors, large envelope, resealable bag

## How to Make the Game:
1. Duplicate three copies of the Halloween Counters, color, laminate, and cut out.
2. Duplicate one copy of the Halloween Graph per child.
3. Store the Halloween Graphs in a large envelope.
4. Store the Counters and several crayons in a resealable bag.

## How to Play the Game:
1. Take one Halloween Graph out of the envelope
2. Spread the Halloween Counters face up on a table.
3. Sort the Halloween Counters by item.
4. Use a different color crayon to mark the number of each item in the correct column on the graph.
5. When you are finished, give your graph to the teacher, and place the game pieces and the crayons back in the bag.

## Option:
Bind the Halloween Graphs in a classroom graph book.

## Book Link:
• *Ed Emberley's Halloween Drawing Book* by Ed Emberley (Little, Brown, 1980). Children will enjoy following these creative drawing tips.

# Halloween Things

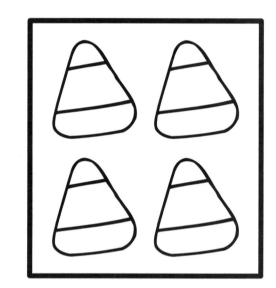

# Halloween Graph

| | | | | | |
|---|---|---|---|---|---|
| 15 | | | | | |
| 14 | | | | | |
| 13 | | | | | |
| 12 | | | | | |
| 11 | | | | | |
| 10 | | | | | |
| 9 | | | | | |
| 8 | | | | | |
| 7 | | | | | |
| 6 | | | | | |
| 5 | | | | | |
| 4 | | | | | |
| 3 | | | | | |
| 2 | | | | | |
| 1 | | | | | |
| | | | | | |

 A to Z Math Games © 1997 Monday Morning Books, Inc.

# HAMBURGER GAME

## Objective:
- Children will match numerals, number words, and sets.

## Materials:
Hamburger Patterns (p. 45), construction paper (red, tan, and brown), scissors, permanent marker, resealable bag

## How to Make the Game:
1. Duplicate the Hamburger Patterns to use as templates.
2. Use the templates to cut out ten red tomatoes, ten tan hamburger buns, and ten brown hamburgers.
3. Write the numbers from 1 to 10 on the hamburger buns. Write the words "one" to "ten" on the hamburgers. Draw sets of dots from one to ten on the tomatoes.
4. Laminate all game pieces and cut them out.
5. Store the game pieces in a resealable bag.

## How to Play the Game:
1. Take the game pieces out of the bag and place them with the numbers, dots, or words face-up on a table.
2. Build hamburgers by correctly matching the buns, tomatoes, and hamburgers. For example, match the tomato with one dot with the hamburger that has the word "one" with the bun that has the number 1.
3. When you have built ten hamburgers, have your teacher check your work. Then place all game pieces back in the bag.

## Book Link:
- *Cloudy with a Chance of Meatballs* by Judi Barrett (Atheneum, 1978).

In the town of Chewandswallow, meals come down from the sky.

# Hamburger Patterns

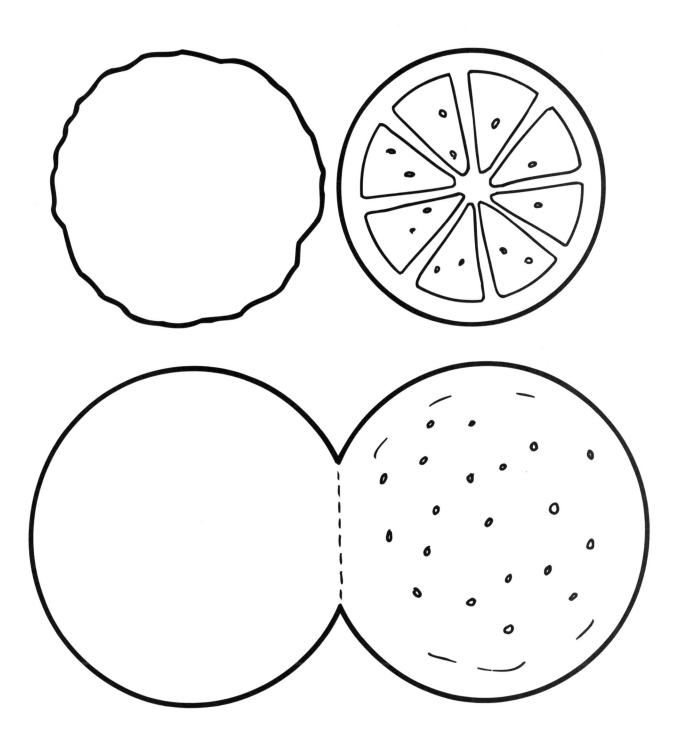

# HOT DOG GAME

## Objective:
- Children will match dots to the correct numerals.

## Materials:
Hot Dogs and Buns (p. 47), crayons or markers, scissors, resealable bag

## How to Make the Game:
1. Duplicate the Hot Dogs and Buns, color, laminate, and cut out.
2. Store the game pieces in a resealable bag.

## How to Play the Game:
1. Spread the game pieces face up on a table, and look at the number on each Hot Dog and the dots on each Bun.
2. Match the correct Hot Dog to each Bun.
3. When you are finished, have your teacher check your work. Then place all game pieces back in the bag.

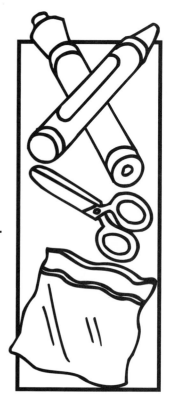

## Book Link:
- *Hotter Than a Hot Dog* by Stephanie Calmenson (Little, Brown, 1994).

A little girl and her grandmother escape the city on a hot summer day by going to the beach.

# Hot Dogs and Buns

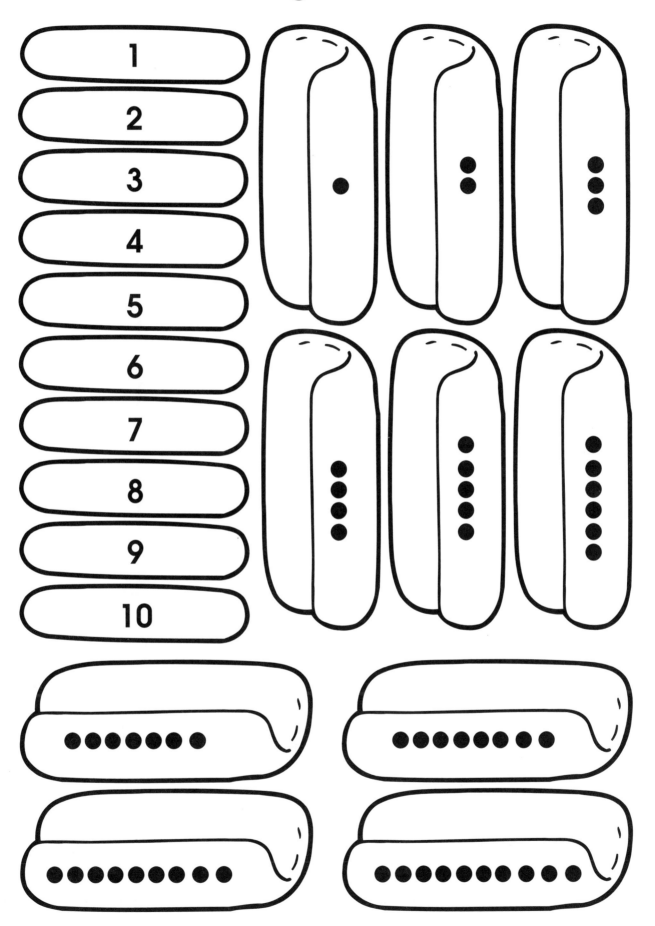

A to Z Math Games © 1997 Monday Morning Books, Inc.

# ICE CREAM MATCH

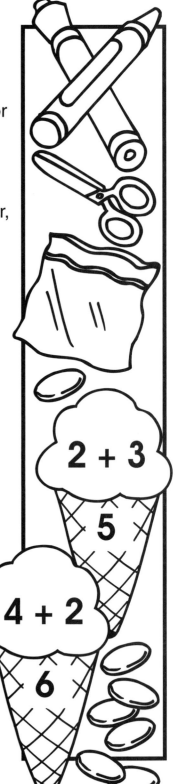

## Objective:
• Children will solve simple addition problems using counters.

## Materials:
Ice Cream Cones (p. 49), Ice Cream Scoops (p. 50), crayons or markers, scissors, counters, resealable bag

## How to Make the Game:
1. Duplicate the Ice Cream Cones and Ice Cream Scoops, color, laminate and cut out.
2. Store all game pieces and counters in the resealable bag.

## How to Play the Game:
1. Take the game pieces out of the bag and spread them face up on a table.
2. Solve the addition problem on each Ice Cream Scoop. (Use counters to help answer the problems.)
3. Match the Ice Cream Cone that has the correct answer to each Ice Cream Scoop.
4. When you are finished, have your teacher check your work. Then place all game pieces back in the bag.

## Option:
Write the answer on the back of each Ice Cream Scoop for self-checking.

## Book Link:
• *Yummers!* by James Marshall (Houghton Mifflin, 1973). Emily the pig tries unsuccessfully to lose weight by exercising.

# Ice Cream Cones

A to Z Math Games © 1997 Monday Morning Books, Inc.

# Ice Cream Scoops

2 + 3

4 + 2

4 + 4

5 + 2

5 + 4

3 + 7

# INCH MEASURING

## Objective:

- Children will practice estimating and measuring in inches.

## Materials:

Measuring Sheet (p. 52), two rulers, large envelope, book, child's scissors, pen, pencil, crayon, bottle of glue

## How to Make the Game:

1. Duplicate one copy of the Measuring Sheet per child.
2. Store the Measuring Sheets in the large envelope.
3. Place the ruler and the items to be measured (second ruler, book, child's scissors, pen, pencil, crayon, and bottle of glue) in the math basket.

## How to Play the Game:

1. Take a Measuring Sheet out of the large envelope.
2. Look at the items to be measured on the sheet, and guess how many inches long each item is. Write this number in the first box on the Measuring Sheet.
3. Use the ruler to measure each item and write the length in the second box on the Measuring Sheet.
4. Compare your guesses with the correct answers.

## Book Link:

- *Inch by Inch: The Garden Song* by David Mallett (HarperCollins, 1995).

A child grows a garden with the help of the rain and the earth.

A to Z Math Games © 1997 Monday Morning Books, Inc.

# Measuring Sheet

Name _____

| Things | Take a guess! | Measure it! |
|---|---|---|
| ruler | | |
| book | | |
| scissors | | |
| pen | | |
| pencil | | |
| crayon | | |
| glue | | |

# JAR LIDS

## Objective:
• Children will practice estimating, counting, and ordering.

## Materials:
Estimation Sheet (p. 54), five different-sized jar lids, permanent marker, crayon, counters (beans, beads, or buttons), large envelope, resealable bag

## How to Make the Game:
1. Duplicate one copy of the Estimation Sheet per child.
2. Store the Estimation Sheets in the large envelope.
3. Use the permanent marker to number the inside of each jar lid (from 1 to 5).
4. Store the jar lids, crayon, and counters in the resealable bag.

## How to Play the Game:
1. Take one Estimation Sheet out of the envelope.
2. Take the crayon and jar lids out of the bag.
3. Guess how many counters the first jar lid might hold.
4. Write the number in the first square of the "guess" column.
5. Scoop up the counters with the jar lid and then count how many are in the scoop.
6. Write this number in the first square of the "answer" column.
7. Continue to guess, count, and record for each jar lid.
8. When you are finished, place the all game pieces back in the bag. Give your Estimation Sheet to the teacher.

## Book Link:
• *Uno, Dos, Tres = One, Two, Three* by Pat Mora, illustrated by Barbara Lavalee (Clarion, 1996). Rhyming text presents numbers in English and Spanish.

| Guess<br>? ? ? ? ? ? ? ? ? | Answer |
|---|---|
| 10 | 6 |
| | |
| | |
| | |

# Estimation Sheet

| Guess<br>? ? ? ? ? ? ? ? ? | Answer |
| --- | --- |
|  |  |
|  |  |
|  |  |
|  |  |
|  |  |

# JAGUARS' SPOTS

## Objective:
• Children will practice matching sets to numbers.

## Materials:
Jaguars (p. 56), crayon, scissors, counters, resealable bag, large envelope

## How to Make the Game:
1. Duplicate one copy of the Jaguars per child.
2. Store the Jaguars in the large envelope.
3. Store the counters and the crayon in the resealable bag.

## How to Play the Game:
1. Take one sheet of Jaguars out of the envelope.
2. Count the number of spots on each Jaguar. (Circle the spots as you count them to help you keep track.)
3. Place the matching number of counters on top of each Jaguar.
4. When you are finished, give your teacher your Jaguars. Then place the counters and the crayon back in the bag.

## Option:
Duplicate the Jaguars, color, laminate, and cut out. Have the children match the number of counters to each Jaguar's spots.

## Book Link:
• *The Jaguar* by Lynn Stone (Rourke, 1989).
This is a nonfiction book about jaguars.

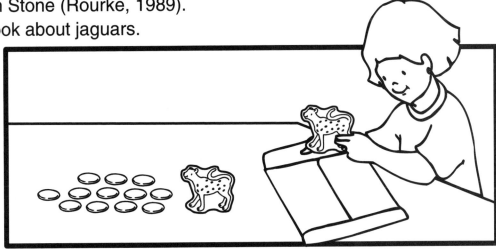

# Jaguars

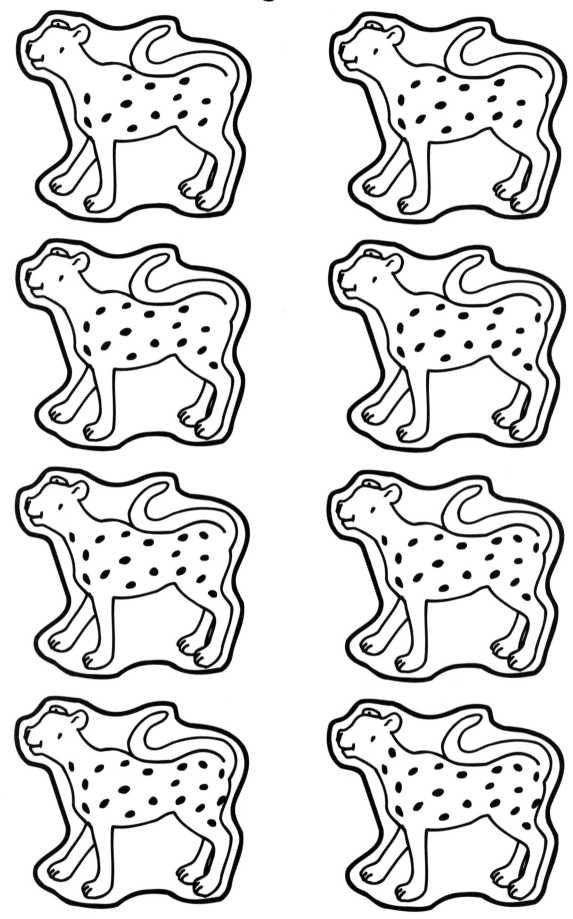

# KINGS' CROWNS

## Objective:
- Children will make sets that match numerals.

## Materials:
Crown (p. 58), Jewels (p. 59), crayons or markers, scissors, large envelope, resealable bag

## How to Make the Game:
1. Make ten copies of the Crown and three copies of the Jewels.
2. Number the Crowns from 1 to 10.
3. Color, laminate, and cut out the Crowns and Jewels.
4. Store the Crowns in the large envelope and the Jewels in the resealable bag.

## How to Play the Game:
1. Take the Crowns out of the envelope and line them up in a row, from 1 to 10.
2. Take the Jewels out of the bag and use them to make number sets for each Crown. For example, place one Jewel on the first Crown, two on the second, and so on.
3. When you are finished, have your teacher check your work. Then put the Crowns in the envelope and the Jewels in the bag.

## Options:
- Decorate the Crowns and Jewels using glitter and glue.
- Let children make sets of beads to place on the Crowns.

## Book Link:
- *King Bidgood's in the Bathtub* by Audrey Wood, illustrated by Don Wood (Harcourt, 1985).
A fun-loving king refuses to get out of the tub.

 A to Z Math Games © 1997 Monday Morning Books, Inc.

# Crown

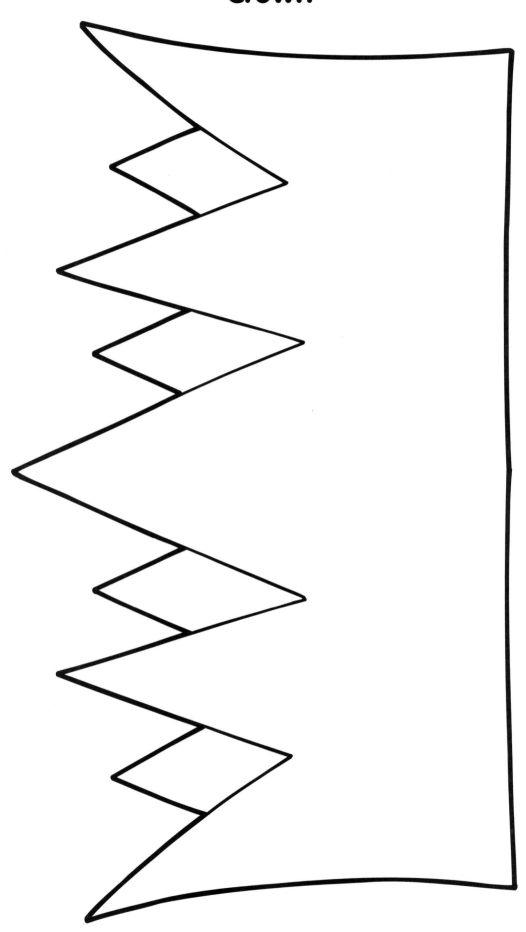

# Jewels

A to Z Math Games © 1997 Monday Morning Books, Inc.

# KITE GAME

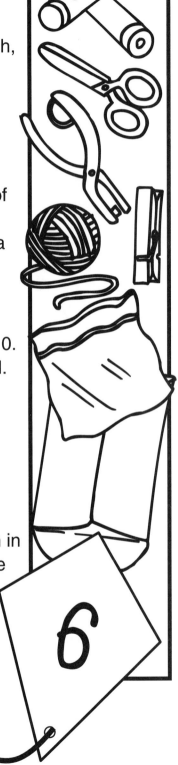

## Objective:
• Children will make sets for numbers from 1 to 10.

## Materials:
Kite (p. 61), 55 clothespins, yarn or ribbon, scissors, hole punch, crayons or markers, large envelope, resealable bag

## How to Make the Game:
1. Make ten copies of the Kite and number them from 1 to 10.
2. Color the Kites, laminate, and cut out.
3. Punch a hole in the bottom of each Kite and attach a piece of yarn or ribbon to represent the tail.
4. Store the Kites in the large envelope and the clothespins in a resealable bag.

## How to Play the Game:
1. Take the Kites out of the envelope, and line them from 1 to 10.
2. Attach the matching number of clothespins to each Kite's tail.
3. When you are finished, have your teacher check your work. Then place the clothespins in the bag and the Kites in the envelope.

## Option:
Use different-colored plastic clothespins, and color the Kites accordingly. (Color Kite #1 blue, and place one blue clothespin in the bag. Color Kite #2 red, and place two red clothespins in the bag, and so on.)

## Book Link:
• *The Sea-Breeze Hotel* by Marcia Vaughan and Patricia Mullins  (Willa Perlman Books, 1992). Kite-flying is the major attraction at the Sea-Breeze Hotel.

# Kite

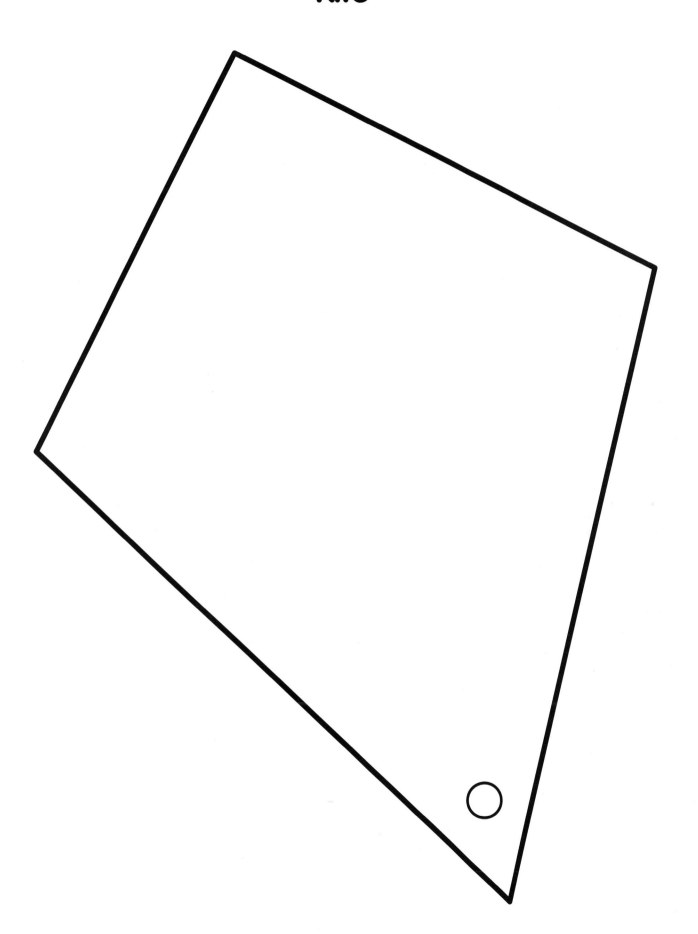

# LADYBUG GAME

## Objective:

• Children will make sets to for the numbers 1 to 15.

## Materials:

Ladybugs (p. 63), permanent marker, crayons or markers, scissors, black beans (or other black counters), resealable bag

## How to Make the Game:

1. Duplicate five copies of the Ladybugs, and number the Ladybugs' shells from 1 to 15.
2. Color the Ladybugs, laminate, and cut out.
3. Store the Ladybugs and the black beans in a resealable bag.

## How to Play the Game:

1. Spread the Ladybugs face up on the floor.
2. Look at the number on the back of each Ladybug.
3. Use the black beans to put the correct number of spots on the back of each Ladybug.
4. When you are finished, have your teacher check your work. Then place the Ladybugs and the counters back in the bag.

## Book Link:

• *Lady Bugatti* by Joyce Maxner, illustrated by Kevin Hawkes (Lothrop, Lee & Shepard, 1991).
Lady Bugatti invites her friends to dinner and to the theater.

# Ladybugs

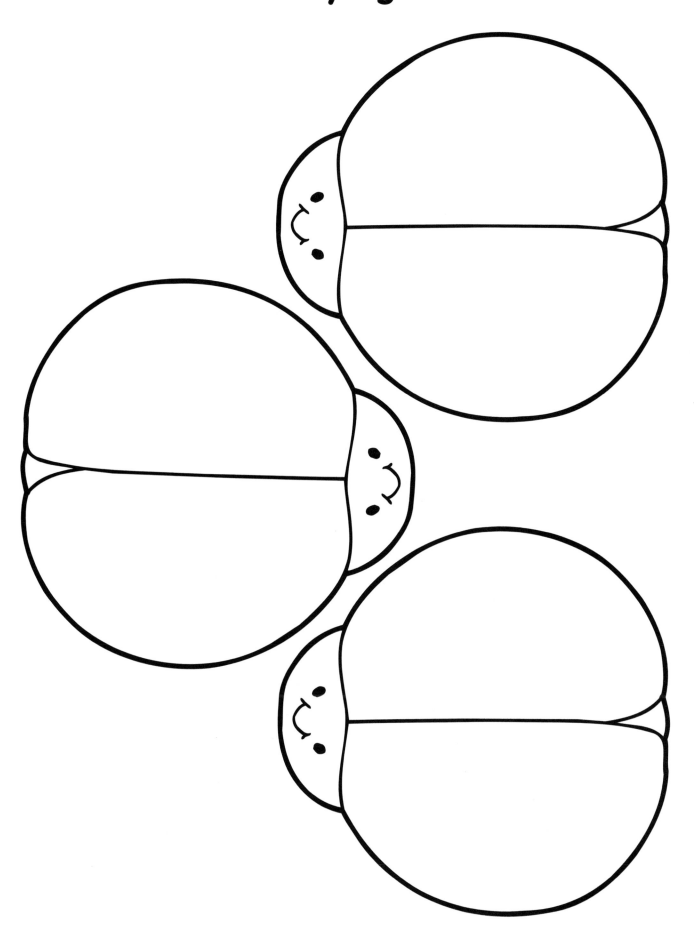

# MORE OR LESS

## Objective:
- Children will compare sets to find out which has more or less.

## Materials:
More or Less Sheet (p. 65), crayon, large envelope

## How to Make the Game:
1. Duplicate one copy of the More or Less Sheet per child.
2. Store the More or Less Sheets in the large envelope along with the crayon.

## How to Play the Game:
1. Take one More or Less Sheet and the crayon out of the envelope.
2. Look at the pictures on the page and count to find out which of the two pictures in each section has more of the item.
3. Circle the pictures that have the most of each item.
4. When you are finished, place the crayon back in the envelope and give your sheet to the teacher.

## Options:
- Have children choose two handfuls of counters and compare the sets, counting to find which handful has more than the other.
- Have children use counters to make sets for the items on the More or Less Sheet.

## Book Link:
- *How Many?* by Debbie MacKinnon (Dial, 1993). This book invites readers to find one child, two candles, three kittens, and other objects up to ten.

# More or Less Sheet

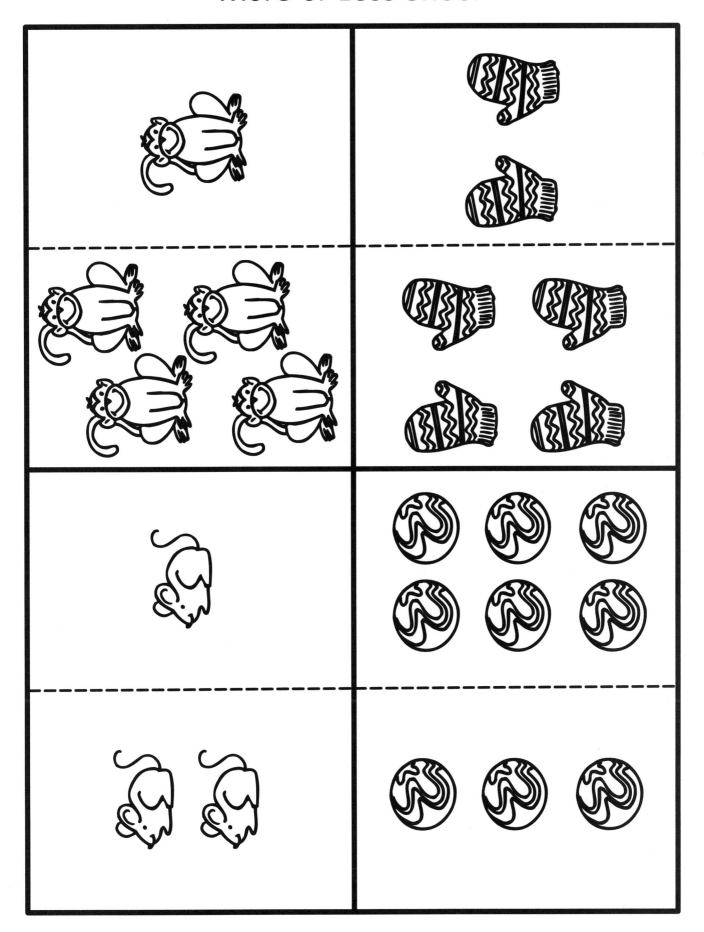

 A to Z Math Games © 1997 Monday Morning Books, Inc.

# METRIC MEASURING

## Objective:
- Children will practice estimating and measuring using metrics.

## Materials:
Measuring Sheet (p. 52), two metric rulers, large envelope, book, child's scissors, pen, pencil, crayon, bottle of glue

## How to Make the Game:
1. Duplicate a copy of the Measuring Sheet for each child.
2. Store the Measuring Sheets in the large envelope.
3. Place the ruler and the items to be measured (second ruler, book, child's scissors, pen, pencil, crayon, and bottle of glue) in the math basket.

## How to Play the Game:
1. Take a Measuring Sheet out of the large envelope.
2. Look at the items to be measured on the sheet.
3. Look at the ruler to see how big a centimeter is. (A centimeter is smaller than an inch.)
4. Look at each item and guess how long it is in centimeters. Write this number in the first box on the Measuring Sheet.
5. Use the ruler to measure each item and write the length in the second box on the Metric Measuring Sheet.
6. Compare your guesses with the correct answers.

## Book Link:
- *One Meter Max* by Ann Segan (Prentice Hall, 1979). Michael Meter, Minnie Millimeter, Cedric Centimeter, and Katie Kilometer explain metric units of measurements.

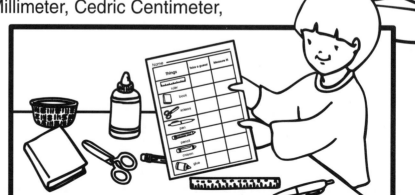

# MUSICAL MATH

## Objective:
● Children will practice repetitive rhythm patterns.

## Materials:
Instrument (maracas, rattle, triangle, bells, and so on), shoe box or other small box, three index cards, marker

## How to Make the Game:
1. Place the instrument in the container.
2. On the index cards, draw different patterns, such as a long line followed by two dots followed by another line, and so on. (See suggested patterns on the bottom of the page.)
3. Store the cards in the box with the instruments.

## How to Play the Game:
1. Take the cards and the instrument out of the box.
2. Look at the pattern on the first card. The long lines represent a long note, and the dots represent short notes.
3. Use the musical instrument to play each pattern. For example, if the card has a long line followed by two dots followed by another long line, you would use the instrument to play a long note, two quick notes, and another long note.
4. Play the pattern on each card.
5. When you are finished, place the cards and the instrument back in the box.

## Suggested Patterns:
● ● _____ ● ● _____ ● ●

_____ ● _____ ● _____

● ● ● _____ ● ● ●

**Note:** This is a slightly noisy activity.

   A to Z Math Games © 1997 Monday Morning Books, Inc.

# MAIL CARRIER

## Objective:
- Children will practice reading three-digit numbers.

## Materials:
Mail Hat (p. 69), Houses (p. 70), Letters (p. 71), blue construction paper, glue or tape, crayons or markers, scissors, resealable bag

## How to Make the Game:
1. Duplicate the Mail Hat onto blue construction paper, color, laminate, and cut out. Fold the brim upward on the dotted line.
2. Make a construction paper band and attach it to the Mail Hat pattern. (Make sure the band is long enough to fit around a child's head.)
3. Tape or glue the ends of the band together to make the hat.
4. Copy the Houses and Letters, color, laminate, and cut out.
5. Store all game pieces in a resealable bag.

## How to Play the Game:
1. Take the Mail Hat out of the bag and put it on.
2. Line the Houses in a row and take the Letters out of the bag.
3. Look at the numbers on the Houses and the Letters.
4. Deliver the mail by matching each Letter to the correct House.
5. When you are finished, have your teacher check your work. Then place all game pieces back in the bag.

## Book Link:
- *The Jolly Postman or Other People's Letters* by Janet and Allan Ahlberg (Little, Brown, 1986).

This inventive book includes real letters in envelopes.

# Mail Hat

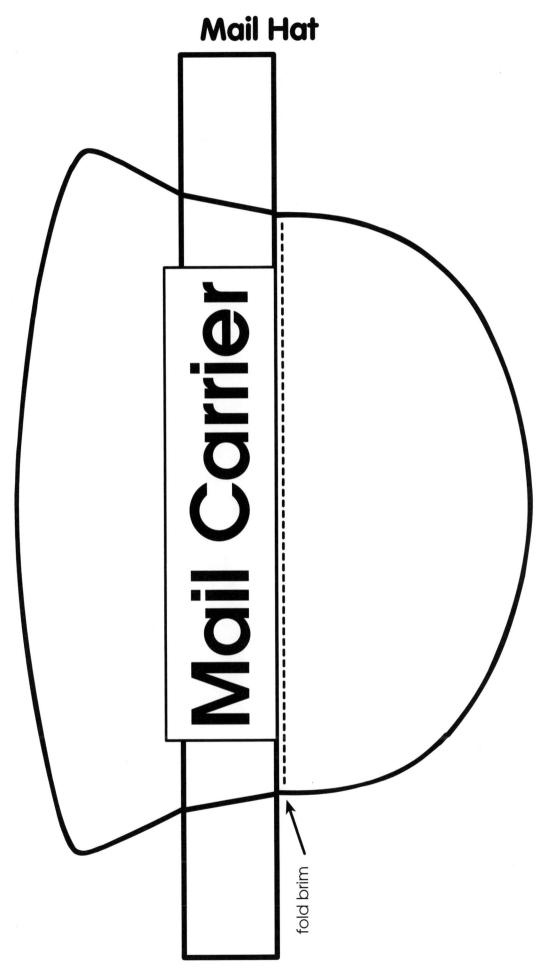

fold brim

Mail Carrier

A to Z Math Games © 1997 Monday Morning Books, Inc.

# Houses

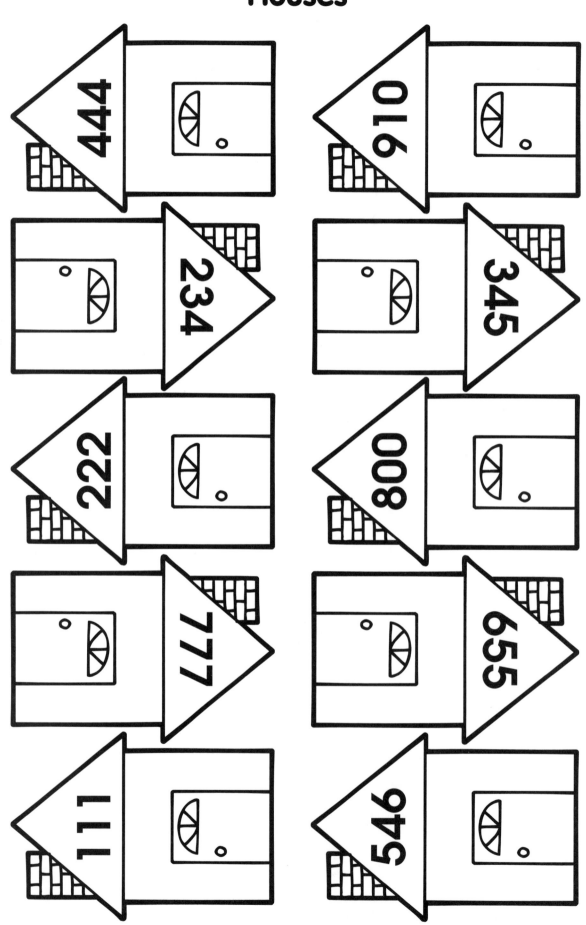

# Letters

| | |
|---|---|
| 111 | 222 |
| 234 | 345 |
| 444 | 546 |
| 655 | 777 |
| 800 | 910 |

# NUMBER NECKLACE

## Objective:
- Children will put the numbers 1 to 10 in the correct order.

## Materials:
Tubular macaroni (10 pieces per set), permanent marker, yarn, scissors, resealable bags

## How to Make one Set:
1. Write one number from 1 to 10 on each piece of macaroni. (Make sure each set has one 1, one 2, and so on, through 10.)
2. Cut a necklace-length piece of yarn for each child.
3. Store the sets in individual bags.

## How to Play the Game:
1. Take the yarn and the pieces of macaroni out of the bag.
2. Line the macaroni pieces in order from 1 to 10.
3. String the pieces of macaroni in order on the piece of yarn.
4. Show the necklace to your teacher when you're finished. Your teacher can help you tie the ends of the necklace together.

## Book Link:
- *Steven Kellogg's Yankee Doodle* by Edward Bangs, illustrated by Steven Kellogg (Parents' Magazine Press, 1976).
This retelling of the famous song includes the sheet music.

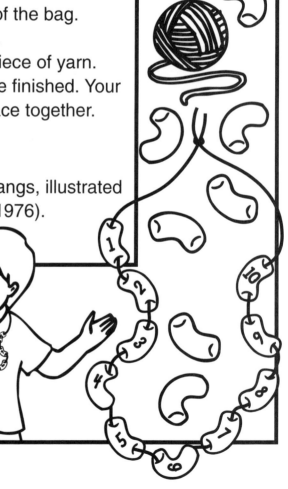

# OLLIE THE OCTOPUS

## Objective:
- Children will practice making sets for the numbers 1 to 8.

## Materials:
Octopus (p. 74), loop cereal (or hole reinforcers), glue, large envelope, resealable bag

## How to Make the Game:
1. Duplicate a copy of the Octopus for each child.
2. Store the Octopi in the large envelope.
3. Store the cereal in a resealable bag.
4. Place the bottle of glue at the Math Center.

## How to Play the Game:
1. Take one copy of the Octopus out of the envelope.
2. Look at the number on each of the Octopus's legs.
3. Glue the correct number of cereal pieces to each one of the Octopus's legs to represent the numbers.
4. When you are finished, give your Octopus to the teacher and place the rest of the materials at the Math Center.

## Option:
Post the completed Octopus art on an "Under the Sea" bulletin board. Let children decorate it using green and blue streamers.

## Book Link:
- *I Was All Thumbs* by Bernard Waber (Houghton Mifflin, 1975).
Legs the octopus is released into the ocean after living his entire life in a lab.

# Octopus

# OSTRICH EGGS

## Objective:
- Children will be able to put numbers in order from 1 to 20.

## Materials:
Ostrich Eggs (p. 76), crayons or markers, scissors, resealable bag

## How to Make the Game:
1. Duplicate 10 copies of the Ostrich Eggs.
2. Number the Ostrich Eggs from 1 to 20.
3. Laminate the Eggs, and cut them out.
4. Store the Ostrich Eggs in a resealable bag.

## How to Play the Game:
1. Take the Ostrich Eggs out of the envelope.
2. Line the Eggs up in order from 1 to 20.
3. Mix the Eggs, and line them up again, this time from 20 to 1.
4. When you are finished, have your teacher check your work. Then place all of the Ostrich Eggs back in the bag.

## Options:
- Share this fact with your children: Ostrich eggs can weigh up to 3.3 lbs.
- Provide weights and a scale for them to see how heavy this is.
- Bring in a chicken's egg for children to weigh in comparison.

## Book Link:
- *Bird Egg Feather Nest* by Maryjo Koch (Stewart, Taori & Chang, 1992).

This is a beautiful book that includes many pictures and facts.

# Ostrich Eggs

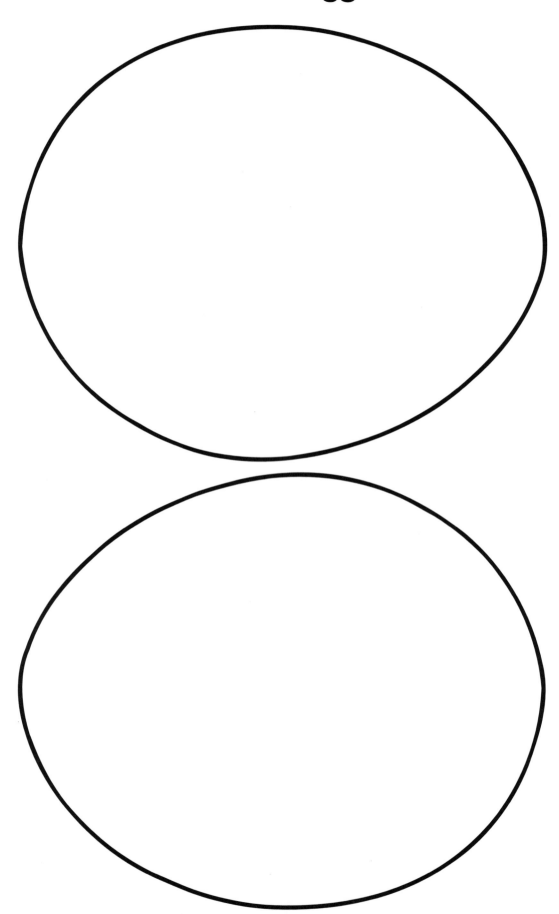

# PUDDING PAINT

## Objective:
- The children will practice writing the numbers 1 to 20.

## Materials:
Instant pudding (plus ingredients to make it), paper plates (one per child), index cards, crayons or markers, spoon

## How to Make the Game:
1. Write the numbers 1 to 20 on the index cards and laminate them.
2. Make the instant pudding. If possible, let the children help you.

## How to Play the Game:
1. Spoon out a small amount of pudding onto a paper plate.
2. Trace the numbers from 1 to 20 in the pudding.
3. If you can't remember what a number looks like, look at the index card.
4. When you are finished, wash your hands and make sure there is no pudding on the table.

**Note:** This activity should be done as a class.

## Options:
- Make enough pudding to serve it as a snack after children draw numbers.
- Use finger paint instead of pudding.

## Book Link:
- *Roly-Poly Pudding* by Beatrix Potter (Warne, 1908).

This is the tale of Samuel Whiskers, or the roly-poly pudding.

# PATTERNS

## Objective:
- Children will make their own patterns.

## Materials:
Colored loop-style cereal (or colored macaroni), straws, glue, scissors, index cards, necklace-sized lengths of yarn (one per child), two resealable bags

## How to Make the Game:
1. Cut the straws into one-inch pieces.
2. Make several sample patterns by gluing the cereal and straw pieces to the index cards.
3. Store the cereal, straws, and yarn in one resealable bag.
4. Store the sample patterns in a separate resealable bag.

## How to Play the Game:
1. Take one piece of yarn out of the bag.
2. Take one of the index cards (with cereal and straws glued on it) out of the bag or make up your own pattern.
3. Use the cereal pieces and straw pieces to make a necklace that follows the same pattern as that on the index card. A pattern means that you will put things in order, for example, two pieces of cereal followed by one straw followed by two pieces of cereal.
4. Show the necklace to your teacher when you're finished. Your teacher can help you tie the ends of the necklace together.
5. Replace the other materials in the bags.

## Book Link:
- *Dinosaur Chase* by Carolyn Otto, illustrated by Thacher Hurd (HarperCollins, 1991). A mother dinosaur reads to her son about a stolen necklace.

# QUILTED NUMBERS

## Objective:
• Children will match sets to numerals from 1 to 9.

## Materials:
Quilt (p. 80), crayons or markers, scissors, large envelope, small counters, resealable bag

## How to Make the Game:
1. Duplicate the Quilt, color, and laminate.
2. Place the counters in a resealable bag.
3. Store the Quilt in a large envelope.

## How to Play the Game:
1. Take the Quilt out of the envelope.
2. Look at the number on each Quilt square.
3. Place the correct number of counters on top of each Quilt square to represent each number.
4. When you are finished, have your teacher check your work. Then place the counters back in the bag and the Quilt back in the envelope.

## Book Link:
• *Luka's Quilt* by Georgia Guback (Greenwillow, 1994).
There is a disagreement over what colors Luka's traditional Hawaiian quilt should be.

 A to Z Math Games © 1997 Monday Morning Books, Inc.

# Quilt

# COUNTING QUAILS

## Objective:

- Children will practice matching sets to numbers.

## Materials:

Mama Quails (p. 82), Baby Quails (p. 83), crayons or markers, scissors, two resealable bags

## How to Make the Game:

1. Duplicate the Mama Quails, color, laminate, and cut out.
2. Duplicate and enlarge the Baby Quails, color, laminate, and cut out.
3. Store the Mama Quails in one resealable bag and the Baby Quails in another.

## How to Play the Game:

1. Take the Mama Quails out of the bag and line them in order.
2. Take the Baby Quails out of the bag.
3. Place the correct group of Baby Quails with each Mama Quail. For example, place the group of three Baby Quails behind the Mama Quail that has the number three on her body.
4. When you are finished, have your teacher check your work. Then place the Mamas in one bag and Babies in the other.

## Book Link:

- *Quail Song* by Valerie Carey (Putnam, 1990). In this retelling of a Pueblo tale, a quail outwits a coyote.

# Mama Quails

# Baby Quails

A to Z Math Games © 1997 Monday Morning Books, Inc.

# RABBITS AND CARROTS

## Objective:
- Children will find solutions to simple addition problems.

## Materials:
Rabbits (p. 85), Carrots (p. 86), crayons or markers, scissors, counters, two resealable bags

## How to Make the Game:
1. Duplicate the Rabbits and Carrots, color, laminate, and cut out.
2. Store the Rabbits and Carrots in one resealable bag and the counters in another.

## How to Play the Game:
1. Take the Rabbits and Carrots out of the bag and line them up in two rows.
2. Solve the addition problem on each Rabbit. (If you need help, use the counters to find the solution for each problem.)
3. Match the correct Carrot to each Rabbit. (Note: Some questions may have the same answer.)
4. When you are finished, have your teacher check your work. Then place the Rabbits and Carrots in one bag and the counters in the other.

## Book Link:
- *The Tale of Peter Rabbit* by Beatrix Potter (Warne, 1902). Peter Rabbit goes on an adventure in Mr. McGregor's garden.

# Rabbits

1 + 4

9 + 1

2 + 2

2 + 4

7 + 2

5 + 5

3 + 4

1 + 2

1 + 1

4 + 4

# Carrots

# RAILROAD CARS

## Objective:

- Children will practice making sets to match numbers.

## Materials:

Engines (p. 88), Railroad Cars (p. 89), crayons or markers, scissors, two resealable bags

## How to Make the Game:

1. Duplicate the Engines, color, laminate, and cut out.
2. Make four copies of the Railroad Cars, color, laminate, and cut out.

## How to Play the Game:

1. Take the Engines out of the bag and line them up.
2. Look at the number on each Engine.
3. Place the correct amount of Railroad Cars behind each Engine to match the numbers written on the Engines. (You will have a few extra Railroad Cars when you are finished.)
4. When you are finished, have your teacher check your work. Then place the Engines back in one bag and the Railroad Cars in another.

## Book Link:

- *The Caboose Who Got Loose* by Bill Peet (Houghton Mifflin, 1971).

Katy Caboose wishes she were almost anything but a caboose. When Katy's bolt breaks, she is free!

A to Z Math Games © 1997 Monday Morning Books, Inc.

# Engines

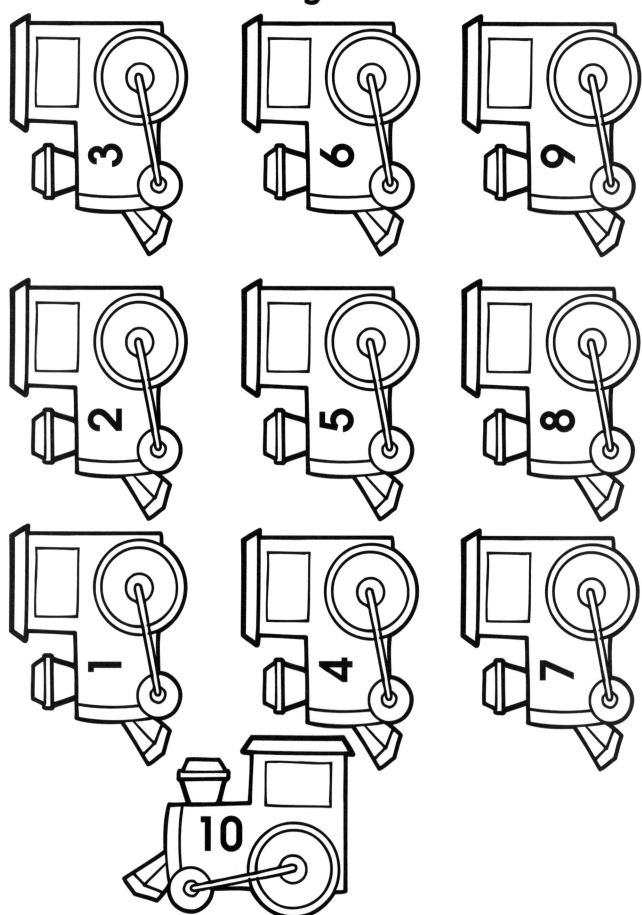

# Railroad Cars

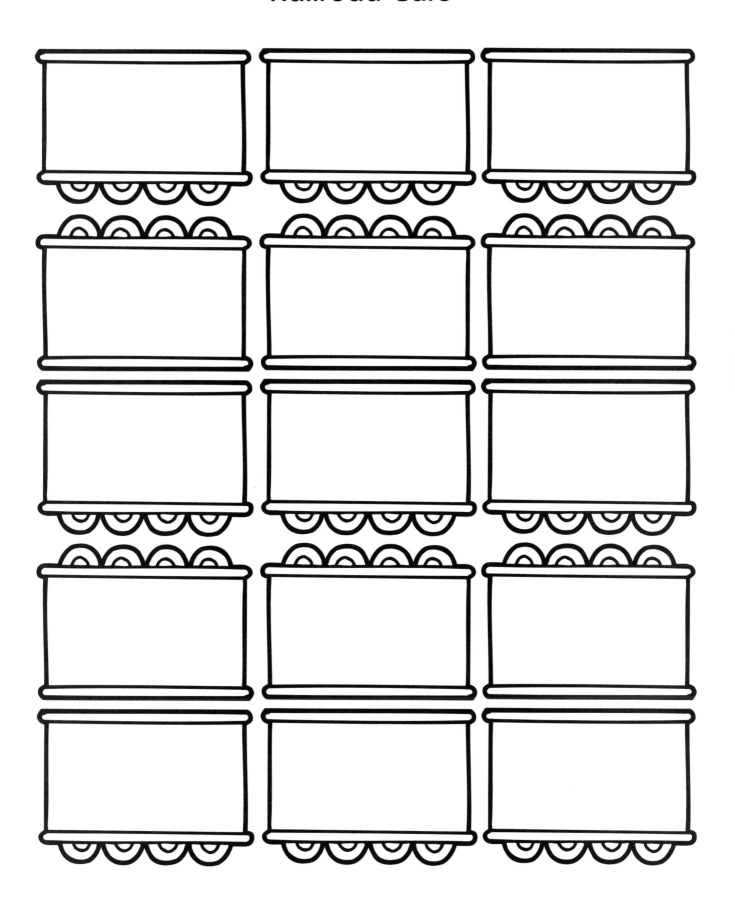

A to Z Math Games © 1997 Monday Morning Books, Inc.

# SPACESHIPS AND STARS

## Objective:
● Children will find solutions to simple subtraction problems.

## Materials:
Spaceships (p. 91), Stars (p. 92), crayons or markers, scissors, counters, two resealable bags

## How to Make the Game:
1. Duplicate the Spaceships and Stars, color, laminate, and cut out.
2. Store the Spaceships and Stars in one bag and the counters in the other.

## How to Play the Game:
1. Take the Spaceships out of the bag and line them face up on a table.
2. Solve the subtraction problem on each Spaceship. (Use the counters to help solve the problems.)
3. Match the Spaceship problems to the Stars that have the correct answers.
4. When you are finished, have your teacher check your work. Then place the Spaceships and Stars back in one bag and the counters in the other.

## Book Link:
● *Nora's Stars* by Satomi Ichikawa (Philomel, 1989). The stars become Nora's toys until the sky cries, and Nora returns them.

# Spaceships

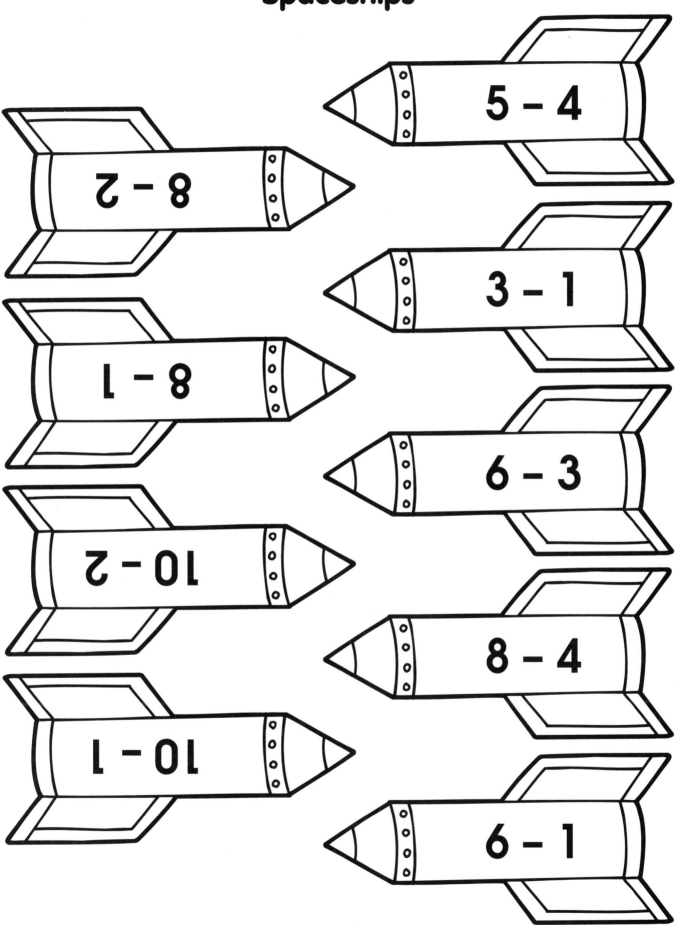

5 – 4

8 – 2

3 – 1

8 – 1

6 – 3

10 – 2

8 – 4

10 – 1

6 – 1

# Stars

# SUBTRACTING SUBS

## Objective:
- Children will find solutions to simple subtraction problems.

## Materials:
Submarines (p. 94), crayons or markers, scissors, counters, hole punch, nine small brads, two resealable bags

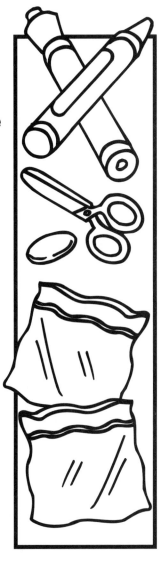

## How to Make the Game:
1. Duplicate the Submarines and Periscopes, color, laminate, and cut out.
2. Punch a hole in each Submarine and Periscope and attach the correct Periscope to the back of each Submarine using brads.
3. Store the Submarines in one resealable bag and the counters in the other.

## How to Play the Game:
1. Spread the Submarines face-up on a table.
2. Look at the subtraction problem on each Submarine.
3. Solve the problems using counters to help you.
4. When you are finished, check your answers by sliding the periscope up on each submarine. Then slide the periscopes back down.
5. Place the Submarines in one bag and the counters in the other.

## Book Link:
- *Pop-Up Numbers* by Ray Marshall and Korky Paul (Dutton, 1994).

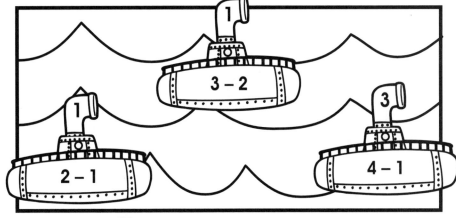

# Submarines

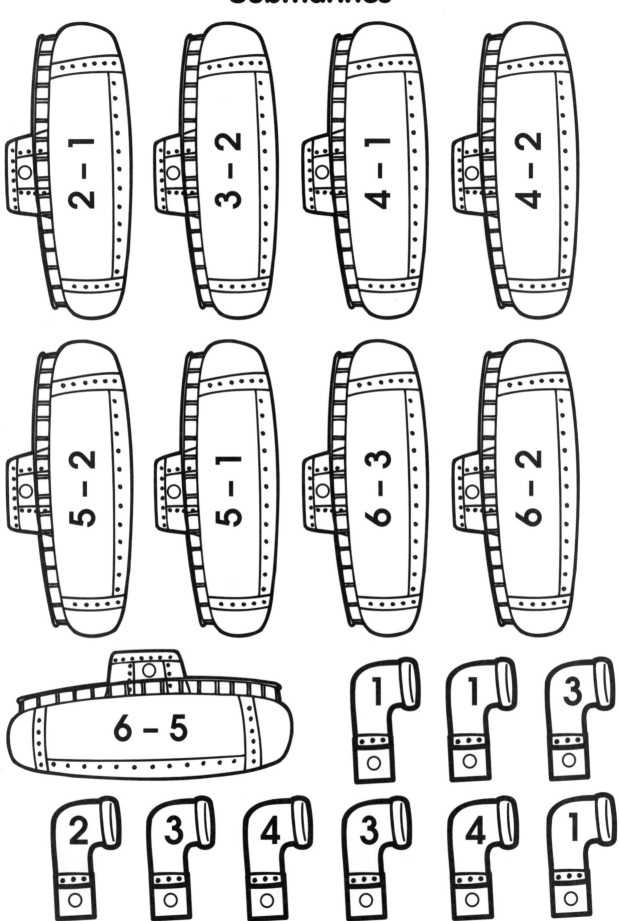

2 − 1

3 − 2

4 − 1

4 − 2

5 − 2

5 − 1

6 − 3

6 − 2

6 − 5

1

1

3

2

3

4

3

4

1

# TINY TEETH

## Objective:
• Children will make sets for the numbers 1 to 15.

## Materials:
Smiley Face (p. 96), unpopped popcorn kernels (120 or more), white spray paint (for adult use only), crayons or markers, scissors, large envelope, resealable bag

## How to Make the Game:
1. Duplicate the Smiley Face 15 times, color, laminate, and cut out.
2. Number the Smiley Faces from 1 to 15.
3. Paint the unpopped popcorn kernels white and let dry.
4. Store the Smiley Faces in the large envelope and the white popcorn kernels in the other.

## How to Use the Game:
1. Take the Smiley Faces out of the bag and line them up in numerical order from 1 to 15.
2. Look at the number on each Smiley Face.
3. Place the correct number of white "teeth" in the mouth of each Smiley Face.
4. When you are finished, have your teacher check your work. Then place the Smiley Faces back in the envelope and the "teeth" in the bag.

## Book Link:
• *The Tooth Fairy* by Peter Collington (Knopf, 1995). A tooth fairy goes to great lengths when a girl loses a tooth.

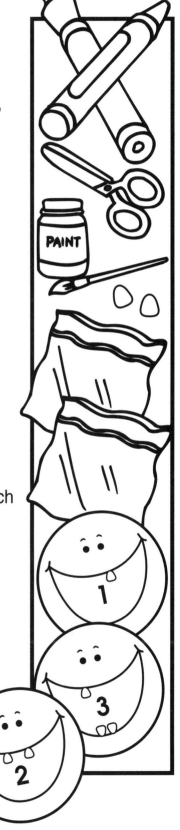

# Smiley Face

# TURTLE TALLYING

## Objective:
- Children will make sets to match numerals from 10 to 20.

## Materials:
Turtles (p. 98), crayons or markers, scissors, unpopped popcorn kernels (at least 165), two resealable bags

## How to Make the Game:
1. Duplicate the Turtles, color, laminate, and cut out. (Enlarge the Turtles if you are using counters larger than popcorn kernels.)
2. Store the Turtles in one resealable bag and the counters in the other.

## How to Play the Game:
1. Take the Turtles out of the bag and spread them face up in numerical order from 10 to 20.
2. Use the counters to make sets on each Turtle's shell. Match the number of counters to the number on each shell.
3. When you are finished, have your teacher check your work. Then place the Turtles in one bag and the counters in the other.

## Book Link:
- *One Was Johnny* by Maurice Sendak (HarperTrophy, 1962).

This counting book includes a turtle, as well as more than ten other characters.

97    A to Z Math Games © 1997 Monday Morning Books, Inc.

# Turtles

# TARANTULAS' BANANAS

## Objective:
- Children will match numbers to sets from 1 to 10.

## Materials:
Tarantulas (p. 100), Bananas (p. 101), crayons or markers, scissors, resealable bag

## How to Make the Game:
1. Duplicate the Tarantulas and Bananas, color, laminate, and cut out.
2. Store all game pieces in a resealable bag.

## How to Play the Game:
1. Take the Tarantulas and Bananas out of the bag and line them face up on a table.
2. Match the number on each Tarantula to the correct group of Bananas. For example, if the number on the Tarantula is eight, match it with the group of eight Bananas.
3. When you are finished, have your teacher check your work. Then place all game pieces back in the bag.

## Option:
A tarantula, with legs extended, can be up to 10 in. (25.4 cm) across. Have children measure this distance and make their own life-size tarantulas from paper and pipe cleaners.

## Book Link:
- *Tarantulas:The Biggest Spiders* by Alexander L. Crosby (Walker, 1981).

This resource is filled with facts and photographs of tarantulas.

# Tarantulas

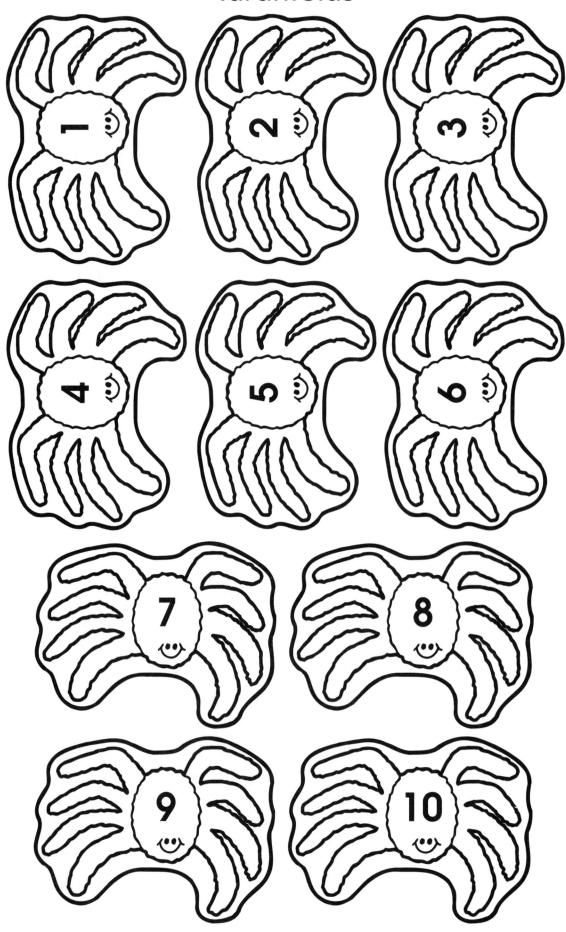

# Bananas

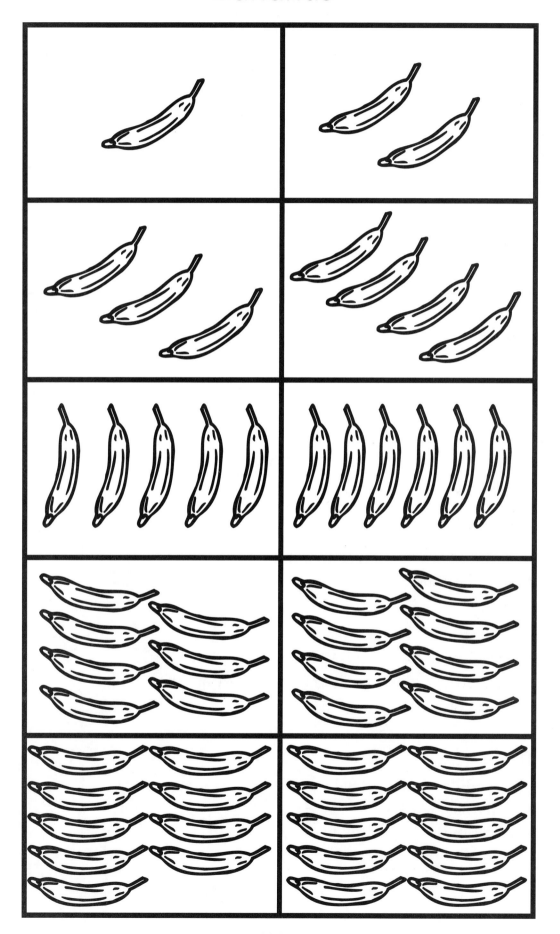

A to Z Math Games © 1997 Monday Morning Books, Inc.

# UMBRELLA GAME

## Objective:
- Children will recognize the numbers 1 to 15 and make sets to represent these numbers.

## Materials:
Umbrellas (p. 103), crayons or markers, scissors, small blue beads (at least 120), two resealable bags

## How to Make the Game:
1. Duplicate the Umbrellas, color, laminate, and cut out.
2. Store the Umbrellas in one resealable bag and the blue beads in the other.

## How to Play the Game:
1. Spread the Umbrellas face up on a table in numerical order from 1 to 15.
2. Make a matching number set of "raindrops" for the number on each Umbrella. (Place the "raindrops" above the umbrellas.)
3. When you are finished, have your teacher check your work. Then place the Umbrellas back in one bag and the blue "raindrops" in the other.

## Option:
Use cocktail umbrellas and number each one from 1 to 15. These are inexpensive and are often available at arts and crafts stores or cooking stores.

## Book Link:
- *The Willow Umbrella* by Christine Widman (Macmillan, 1993).

Two girls discover how much fun it can be to play in the rain.

# Umbrellas

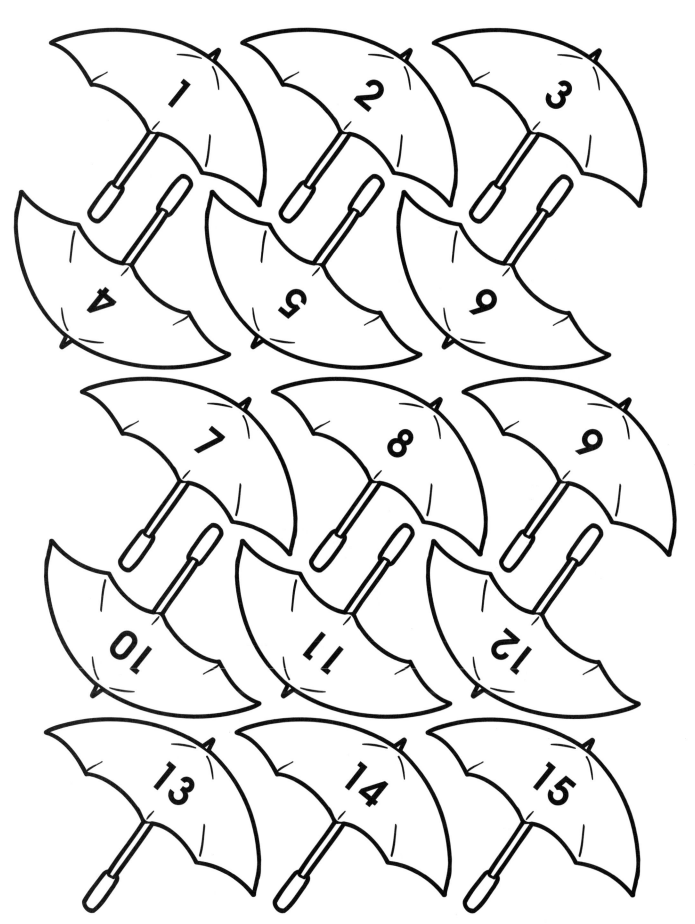

A to Z Math Games © 1997 Monday Morning Books, Inc.

# VALENTINE PUZZLES

## Objective:
- Children will match numbers to corresponding sets of objects.

## Materials:
Valentines (p. 105), crayons or markers, scissors, resealable bag

## How to Make the Game:
1. Duplicate the Valentines, color, laminate, cut out, and cut in half.
2. Store the Valentines in a resealable bag.

## How to Play the Game:
1. Take the game pieces out of the bag.
2. Line the numbered Valentine halves in a row from 1 to 10.
3. Look at the other Valentine halves. Each one has a set representing a number.
4. Match the Valentine set halves with the correct Valentine number halves.
5. When you are finished, have your teacher check your work. Then place all Valentine pieces back in the bag.

## Book Link:
- *My First Valentine's Day Book* by Marian Bennett (Children's Press, 1985).

This easy-to-read nonfiction book tells about making valentines, having parties, and sending cards to friends.

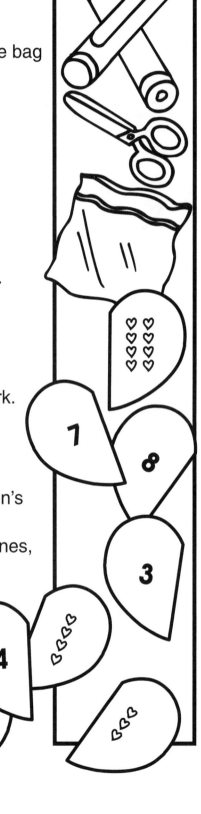

# VALENTINES

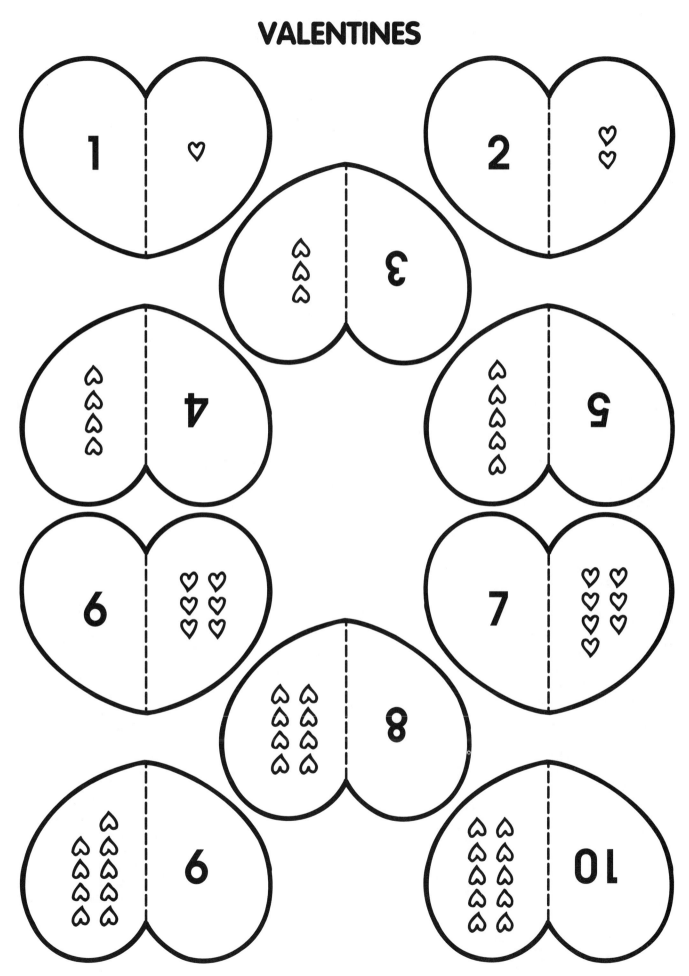

# VALENTINE NUMBERS

## Objective:

- Children will practice writing the numbers from 10 to 20.

## Materials:

Valentine (p. 107), crayons or markers (at least 11 different colors), large envelope, resealable bag

## How to Make the Game:

1. Duplicate a copy of the Valentine for each child.
2. Store the Valentines in the large envelope.
3. Store the crayons or markers in the resealable bag.

## How to Play the Game:

1. Take a Valentine out of the large envelope.
2. Look at the numbers in the different sections of the Valentine.
3. Copy each number over and over to fill each section.
4. When you are finished, give your picture to the teacher and place the crayons or markers back in the bag.

## Option:

Post the completed Valentines on a bulletin board in the classroom. Or let children give these pictures to their friends or relatives on Valentine's Day.

## Book Link:

- *Valentine's Day* by Joyce Kessel (Carolrhoda Books, 1981).

Valentine's Day started more than 2,700 years ago. This nonfiction resource includes many other facts to share.

# Valentine

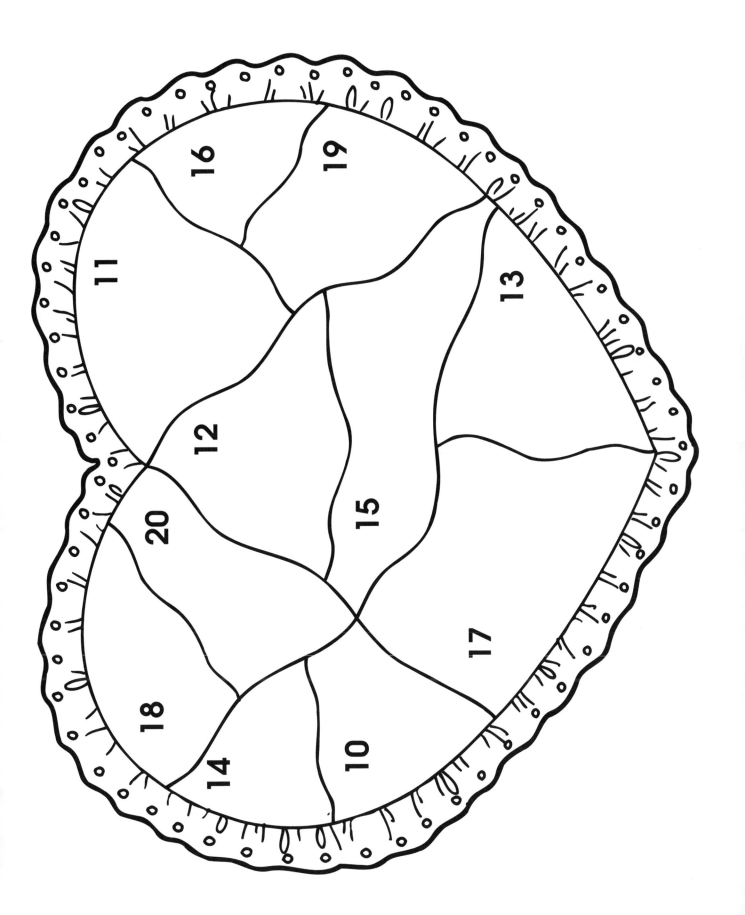

# VAMPIRE BATS

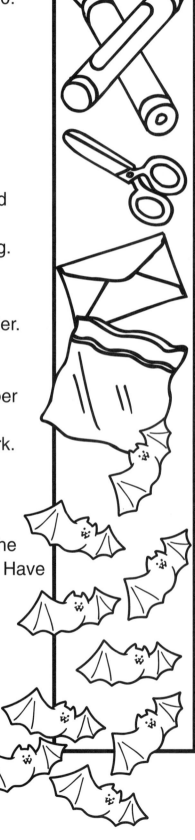

## Objective:
- Children will practice making sets for numbers from 1 to 10.

## Materials:
Vampire Bats (p. 109), 10 envelopes, crayons or markers, scissors, resealable bag

## How to Make the Game:
1. Number the envelopes from 1 to 10.
2. Make four copies of the Vampire Bats, color, laminate, and cut out.
3. Store the Vampire Bats and envelopes in a resealable bag.

## How to Play the Game:
1. Take the envelopes out of the bag and line them up in order. (These envelopes represent caves.)
2. Look at the number on the outside of each cave.
3. In each cave, place the number of bats to equal the number on the outside.
4. When you are finished, have your teacher check your work. Then place all game pieces back in the bag.

## Option:
Duplicate one copy of the Vampire Bats for each child. Let the children glue the bats to sheets of black construction paper. Have the children number the bats from 1 to 15.

## Book Link:
- *The World of Bats* by Virginia Harrison, photographs by Oxford Scientific Films (Gareth Stevens, 1989).
This bat resource includes facts and photographs.

# Vampire Bats

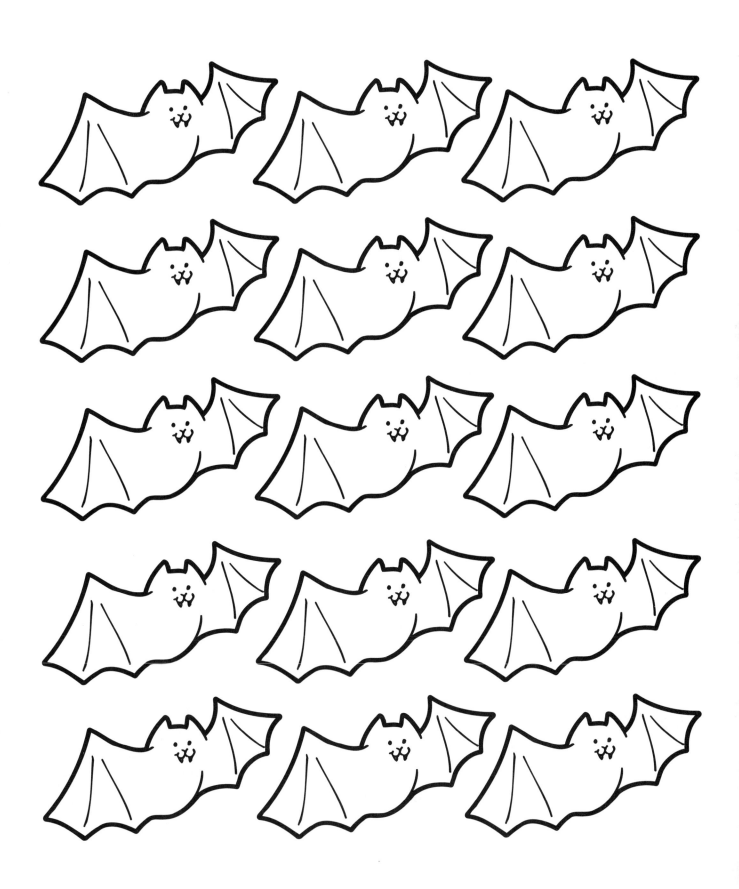

 A to Z Math Games © 1997 Monday Morning Books, Inc.

# VOLCANO MATH

## Objective:

- Children will practice making sets to match numbers.

## Materials:

Volcanoes (p. 111), small red counters, crayons or markers, large envelope, resealable bag

## How to Make the Game:

1. Duplicate and enlarge the Volcanoes, color, and laminate.
2. Store the Volcanoes in a large envelope.
3. Store the red counters in the resealable bag.

## How to Play the Game:

1. Take the sheet of Volcanoes out of the large envelope.
2. Look at the number on each Volcano.
3. Place the correct number of red counters shooting out of each Volcano. For example, on the Volcano numbered 15, place 15 counters shooting out of the top.
4. When you are finished, have your teacher check your work. Then place the Volcanoes back in the envelope and the counters in the resealable bag.

## Options:

- Duplicate a copy of the Volcanoes for each child. Have children place stickers or glue sequins on the Volcanoes.
- Post the pictures on a Volcano Math bulletin board.

## Book Link:

- *Volcanoes - Fire from Below* by Jenny Wood (Gareth Stevens, 1991).

This informative book has text and pictures about volcanoes.

# Volcanoes

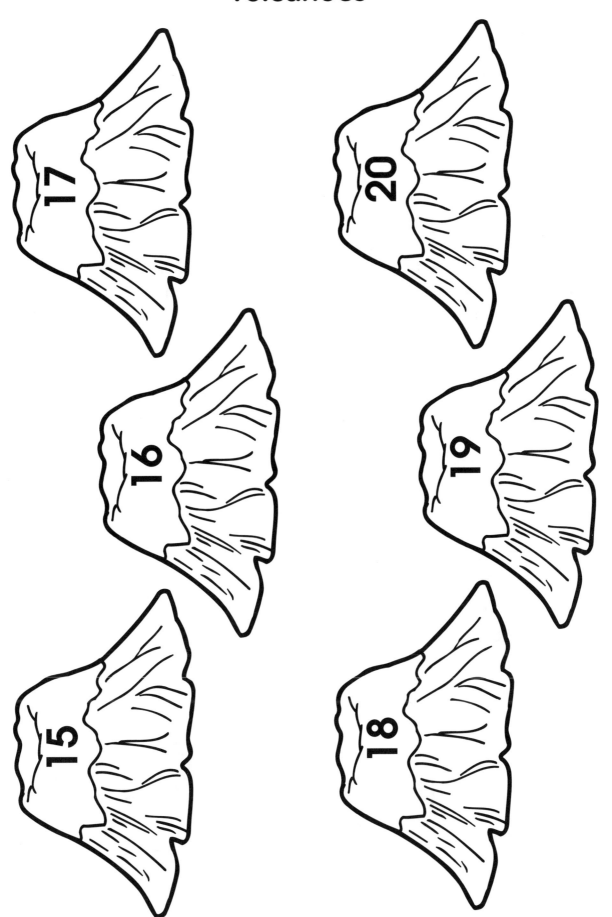

# WHALES' SPOUTS

## Objective:
- Children will practice simple subtraction problems.

## Materials:
Whales (p. 113), Spouts (p. 114), crayons or markers, scissors, counters, two resealable bags

## How to Make the Game:
1. Duplicate the Whales and Spouts, color, laminate, and cut out.
2. Store the Whales and Spouts in one resealable bag and the counters in the other.

## How to Play the Game:
1. Take the Whales out of the bag and line them face up in a row.
2. Use the counters to solve the subtraction problem on each Whale.
3. For each Whale, find the Spout that has the correct answer.
4. When you have matched each Whale with a Spout, have your teacher check your work. Then place all Whales and Spouts back in one bag and all counters in the other.

## Book Link:
- *The Blue Whale* by Kazue Mizumura (Crowell, 1971).
This is a Let's Read-and-Find-Out science book.

# Whales

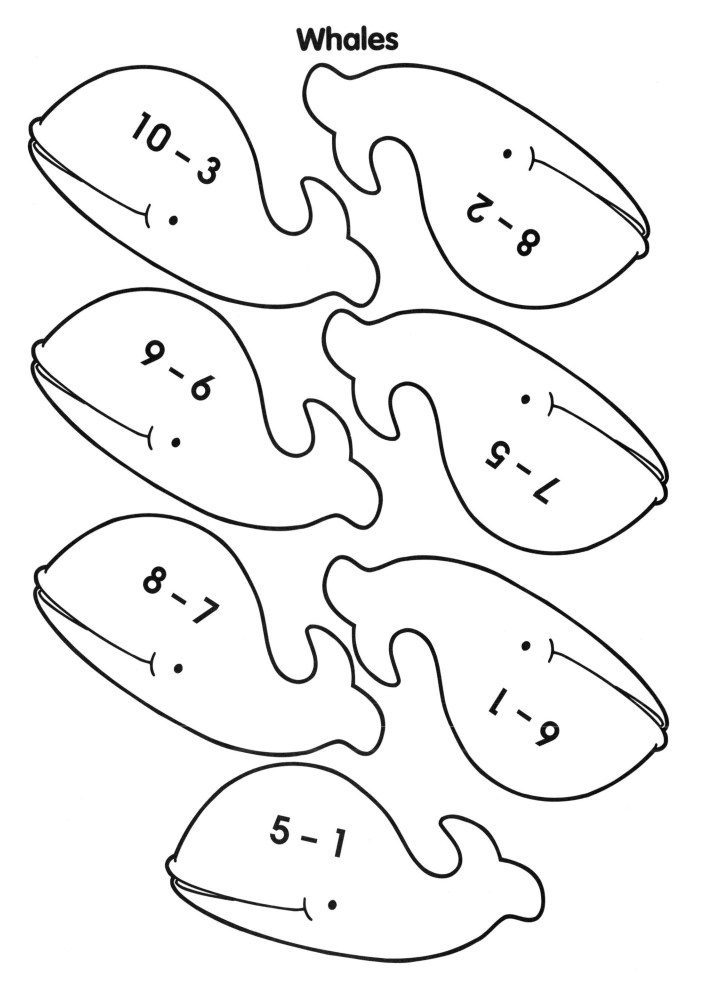

A to Z Math Games © 1997 Monday Morning Books, Inc.

# Spouts

# WATER LILY MATH

## Objective:
- Children will match sets with numbers.

## Materials:
Water Lilies (p. 116), Frogs (p. 117), crayons or markers, scissors, large envelope, resealable bag

## How to Make the Game:
1. Duplicate the Water Lilies, color, and laminate.
2. Duplicate the Frogs, color, laminate, and cut out.
3. Store the Water Lilies in a large envelope and the Frogs in a resealable bag.

## How to Play the Game:
1. Take the Water Lilies out of the envelope.
2. Look at the number on each Water Lily.
3. For each Water Lily, find the Frog with the matching number of dots on its back.
4. When you have matched each Frog and Water Lily, have your teacher check your work. Then place the Water Lilies back in the envelope and the Frogs back in the bag.

## Book Link:
- *Linnea in Monet's Garden* by Christina Bjork, drawings by Lina Anderson (Rabén & Sjögren, 1985).

This adorable book includes reproductions of Monet's work, as well as photographs of his home and garden at Giverny.

# Water Lilies

11

12

13

14

15

16

17

18

19

20

# Frogs

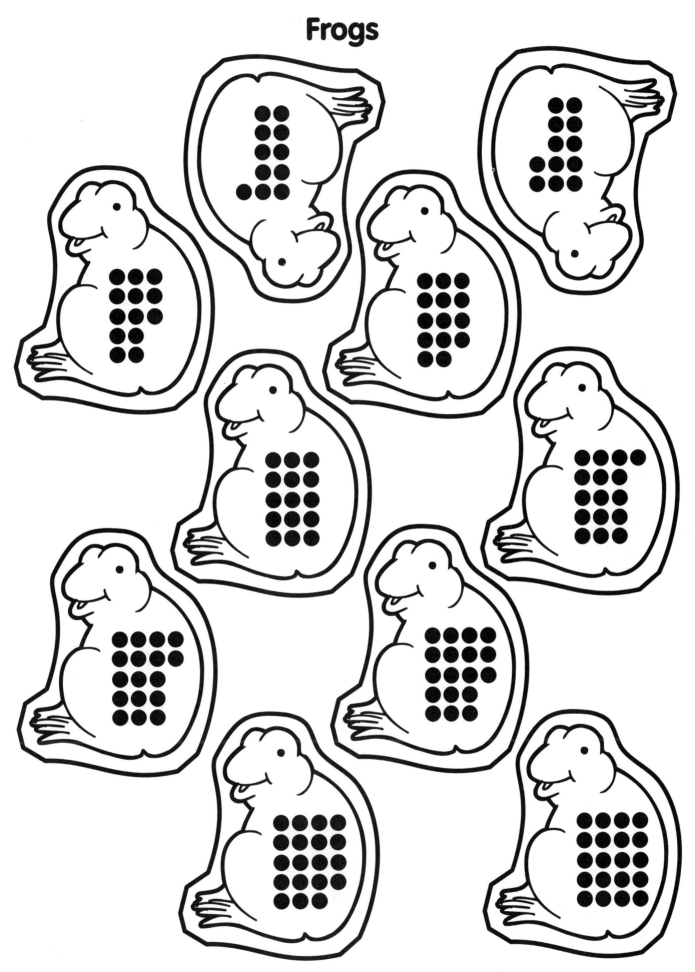

A to Z Math Games © 1997 Monday Morning Books, Inc.

# WHEELS

## Objective:

- Children will practice matching sets to even numbers.

## Materials:

Truck Bodies (p. 119), Wheels (p. 120), crayons or markers, scissors, two envelopes

## How to Make the Game:

1. Duplicate and enlarge the Truck Bodies, color, and laminate.
2. Duplicate two copies of the Wheels, laminate, and cut out.
3. Store the Truck Bodies and the Wheels in the envelopes.

## How to Play the Game:

1. Take the Truck Bodies and Wheels out of the envelopes.
2. Look at the number on each Truck Body.
3. Place the matching number of Wheels on each Truck.
4. When you are finished, have your teacher check your work. Then place the game pieces back in the two envelopes.

## Option:

Duplicate a copy of the Truck Bodies for each child to glue to a sheet of construction paper. Have children draw wheels on the trucks to match the number on each Truck Body.

## Book Link:

- *Things That Go* by Anne Rockwell (Dutton, 1986).
This book is divided into sections, including things that go on the road, on the water, in the air, and more.

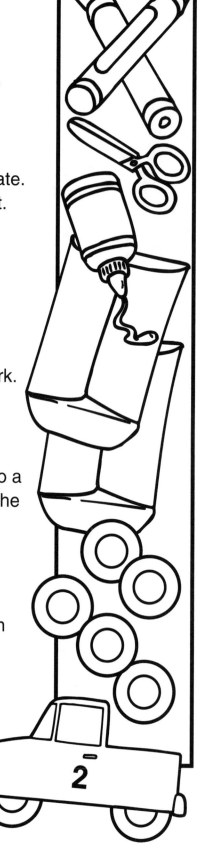

# Truck Bodies

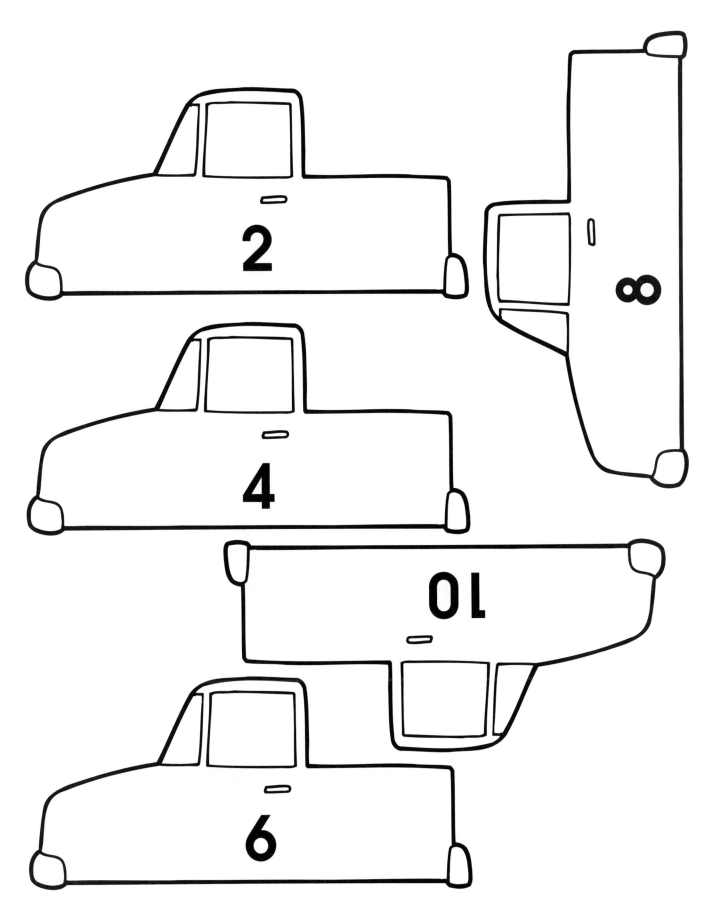

 A to Z Math Games © 1997 Monday Morning Books, Inc.

# Wheels

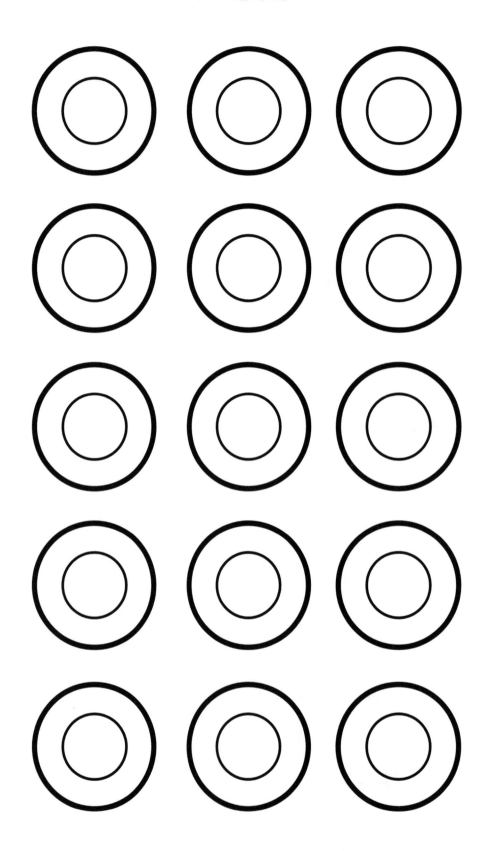

# XYLOPHONE

## Objective:
• Children will practice counting forwards and backwards.

## Materials:
Spinner and Arrow (p. 122), hole punch, brad, crayons or markers, child's xylophone, masking tape, permanent marker

## How to Make the Game:
1. Duplicate the Spinner and Arrow, color, laminate, and cut out.
2. Punch a hole in the center of the Spinner and attach the Arrow using the brad.
3. Use masking tape and a permanent marker to number the xylophone keys in order from 1 to 15.
4. Store the xylophone and Spinner at the Math Center.

## How to Play the Game:
1. Spin the spinner and count from one to the number spun, striking the xylophone keys as you count.
2. Count backward from the number spun to one, striking the keys as you count.
3. Spin the spinner and count forwards and backwards at least five times before putting the game away.

## Book Link:
• *Zin! Zin! Zin! A Violin* by Lloyd Moss, illustrated by Marjorie Priceman (Simon & Schuster, 1995).
Ten musical instruments play in a musical performance.

# Spinner and Arrow

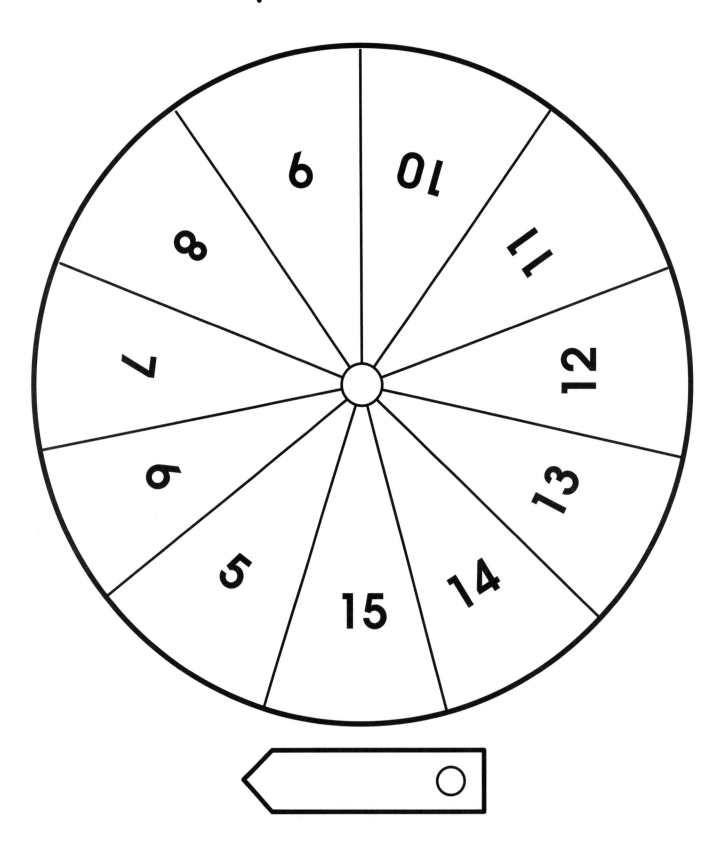

# YOLK ADDITION

## Objective:
- Children will practice solving simple addition problems.

## Materials:
Frying Pans (p. 124), Fried Eggs (p. 125), crayons or markers, scissors, counters, two resealable bags

## How to Make the Game:
1. Duplicate the Frying Pans and Fried Eggs, color, laminate and cut out.
2. Store the Frying Pans and Fried Eggs in one resealable bag and the counters in another.

## How to Play the Game:
1. Take the Frying Pans out of the bag and line them up.
2. Use the counters to help you solve the addition problems on each Frying Pan.
3. Match each Frying Pan with the Egg that has the correct answer on it.
4. When you are finished, have your teacher check your work. Then place all Frying Pans and Eggs back in one bag and the counters in the other.

## Book Link:
- *The Wonderful Egg* by G. Warren Schloat, Jr. (Scribner's, 1952).

This black and white photograph book answers almost every possible question about chickens and eggs!

 A to Z Math Games © 1997 Monday Morning Books, Inc.

# Frying Pans

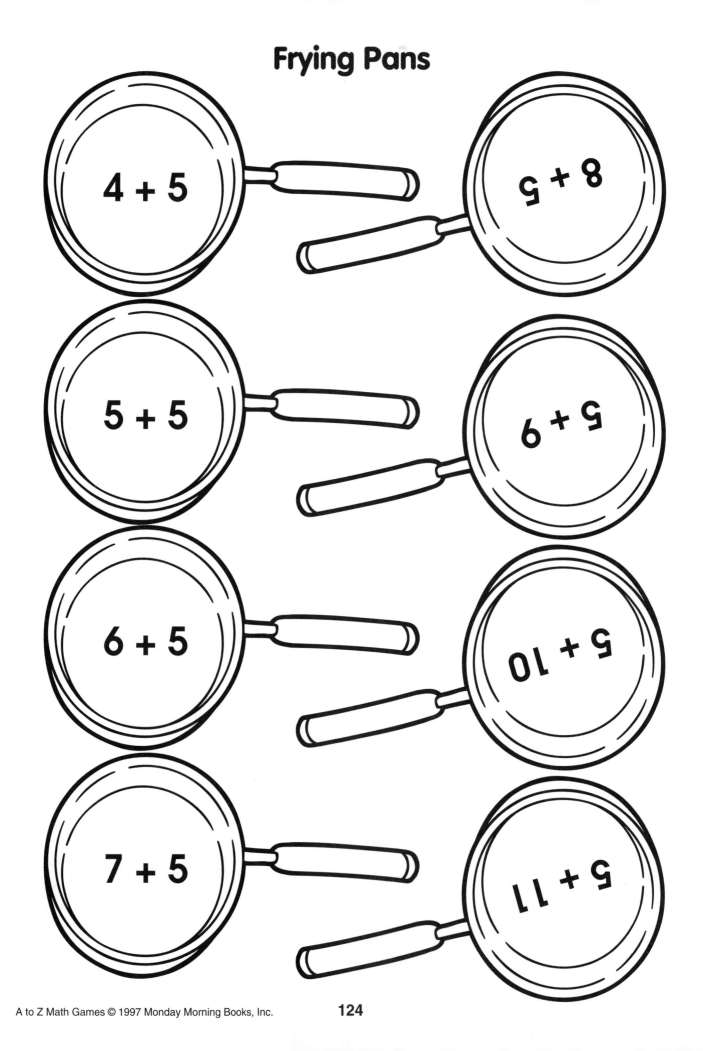

4 + 5

5 + 8

5 + 5

6 + 5

6 + 5

5 + 10

7 + 5

5 + 11

# Fried Eggs

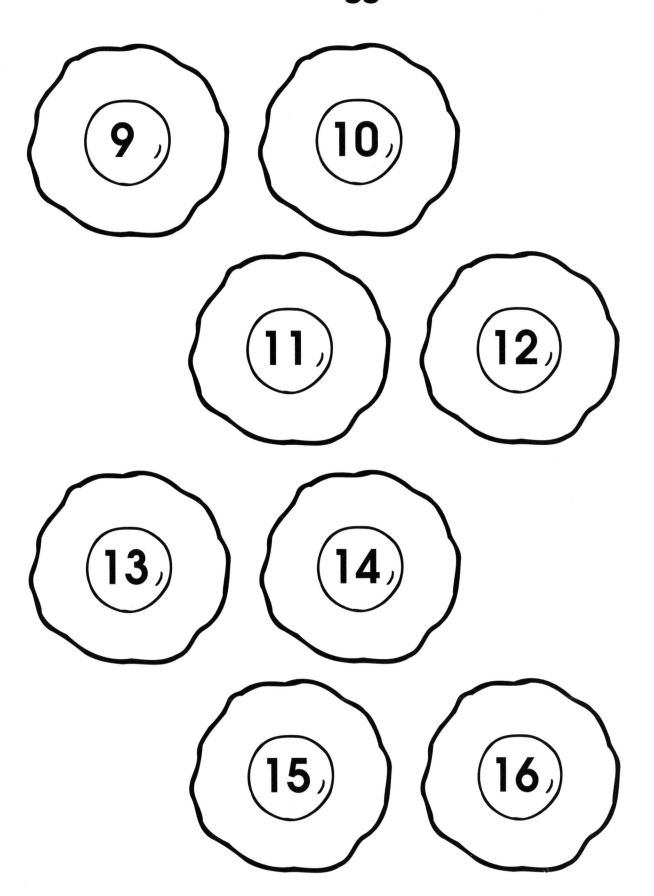

A to Z Math Games © 1997 Monday Morning Books, Inc.

# ZOO ANIMALS

## Objective:
- Children will practice pairing matching numbers.

## Materials:
Animals (pp. 127-128), crayons or markers, scissors, ten empty berry baskets, index cards, tape, resealable bag

## How to Make the Game:
1. Duplicate the Animals, color, laminate, and cut out.
2. Label each berry basket from one to ten using the index cards, markers, and tape.
3. Stack the berry baskets to store.
4. Store the Animals in a resealable bag and place the bag in the top berry basket.

## How to Play the Game:
1. Take the Animals out of the bag and look at the number written on each one.
2. Line the animals in a row from 1 to 10.
3. Line the berry basket "cages" in a row from one to ten.
4. Place the correct animal in a "cage" by matching the numbers on the animals to the numbers on the "cages."
5. When you are finished, place the Animals back in the bag. Stack the baskets. Place the bag of Animals in the top basket.

## Book Links:
- *If I Ran the Zoo* by Dr. Seuss (Random House, 1950).
- *Zachary Goes to the Zoo* by Jill Krementz (Random House, 1986).
- *Zoo* by Anthony Browne (Random House, 1992).

# Animals

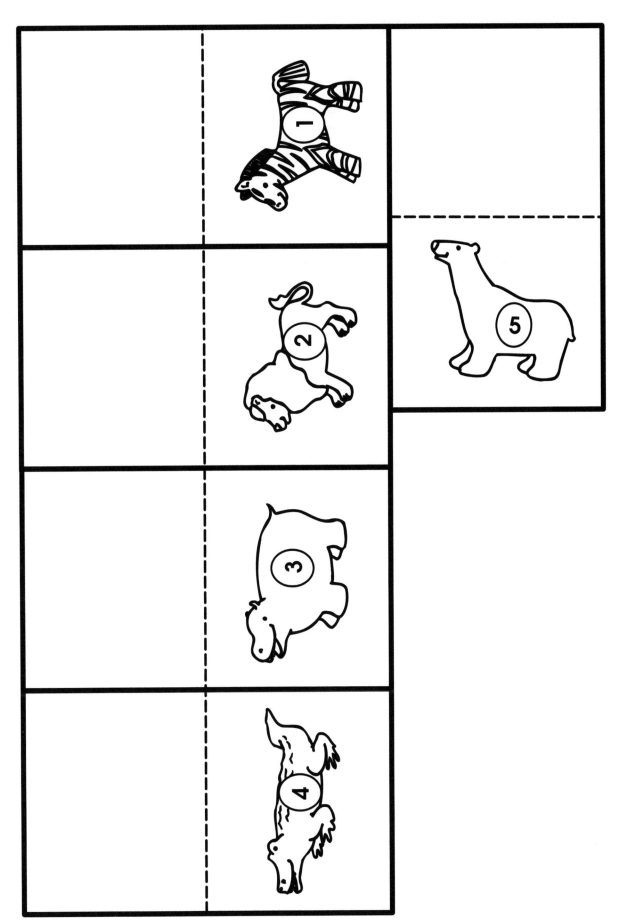

# Animals